Biodiversity and domestication of yams in West Africa

Traditional practices leading to *Dioscorea rotundata* Poir.

Roland Dumont, Alexandre Dansi,
Philippe Vernier, Jeanne Zoundjihèkpon

Edited by David Manley

CIRAD

IPGRI

First published by CIRAD and IPGRI in French in 2005 under the title
*Biodiversité et domestication des ignames en Afrique de l'Ouest - Pratiques
traditionnelles conduisant à Dioscorea rotundata Poir.*

© CEMAGREF, CIRAD, IFREMER, INRA, IPGRI, 2006
ISBN (CIRAD): 2-87614-632-0,
ISBN-13 (IPGRI): 978-92-9043-704-8, ISBN-10 (IPGRI): 92-9043-704-9
ISSN: 1773-7923

CIRAD

The Agricultural Research Centre for International Development, CIRAD, is a French agricultural research centre working for development in developing countries and the French overseas regions. Most of its research is conducted in partnership. It employs 1,820 people, including 1,050 senior staff members, and has an annual operating budget of €200 million.

IPGRI

International Plant Genetic Resources Institute, IPGRI, is an international research institute with a mandate to advance the conservation and use of genetic diversity for the well-being of present and future generations. It is a Centre of the Consultative Group on International Agricultural Research (CGIAR). Founded in 1974 IPGRI has a staff of about 300, in 22 offices around the world. Its budget in 2004 was US$35 million.

Tribute

This book is dedicated to the memory of I.H. Burkill († 1965), D.G. Coursey († 1983) and J. Miège († 1993), our illustrious predecessors in the world of Dioscoreaceae. Their observational and intuitional talents generated the basic knowledge required to study African yams, thus paving the way to a fascinating and disconcerting plant kingdom that never ceases to stimulate scientific curiosity.

One major concern of these three botanists was to gain insight into the origin of *Dioscorea rotundata* yams. The wild parents were identified presumptively but the technical sequences leading to their cropping remained unexplained. The results of several recent studies now provide sufficiently solid arguments for the re-examination of these questions. Various interpretations and hypotheses put forward in this book are still open to debate and require further research.

This in-depth study was undertaken by French-speaking scientists, but the bibliographical references highlight the substantial contribution of the English scientific literature. The generally high quality input of African researchers on both sides of the linguistic divide has also been considerable.

Contents

Acknowledgements

The authors sincerely thank the following people for agreeing to review the manuscript and for their relevant suggestions and advice:

Annette Hladick, General Ecology Laboratory, Muséum national d'histoire naturelle (MNHN), Centre national de la recherche scientifique (CNRS)

Jean-Louis Pham, Diversity and Genomes of Cultivated Plants (Joint Research Unit 1097), Institut de recherche pour le développement (IRD)

Raymond Vodouhé, Sub-Regional Office for West Africa, International Plant Genetic Resources Institute (IPGRI)

Vincent Lebot, Breeding for Vegetatively Propagated Crops (Research Unit 75), Centre de coopération internationale en recherche agronomique pour le développement (CIRAD).

Foreword

The domestication of wild yams is still common practice in West Africa. This phenomenon offers one of the few remaining opportunities to gain insight into how farmers use their empirical knowledge to tap the genetic resources of wild plants and create products suitable for agriculture. Strangely enough, until recently, yam agronomists and breeders have not focused much attention on, or have completely ignored, this agricultural biodiversity generating and organizing process.

The present book aims to fill the gap by pooling existing knowledge on the subject. The prospects for scientific progress in this original field are considerable, at a time when scientists are becoming aware of the potential for technical progress and adaptation to environmental change based on farmers' knowledge and practices relating to genetic resource management.

It deliberately deals only with domestication leading to *Dioscorea rotundata* yams, which by far represent the most widely cultivated type in West Africa and throughout the world. However, the taxonomy and botanical identity of this yam and its wild parents must be clarified before domestication is discussed. A large initial section of this book is therefore devoted to the biodiversity of *D. rotundata* yams and the wild forms from which they derive.

The opening chapter defines and characterizes *D. rotundata* yams in terms of their phyletic relations with the *D. cayenensis* species and also on the basis of botanical, agronomic, technical and genetic criteria.

The following chapter focuses on *D. abyssinica* and *D. praehensilis*, i.e. wild yams used by 'domesticator' farmers to create *D. rotundata* yams. It highlights their relations with different ecosystems and their diversity, differences and similarities.

The final chapter of this first section presents and discusses different phenomena that could modify the variability of these wild yams and make them suitable for domestication.

Domestication is examined in detail only after addressing these different topics. Chapter 5 analyzes the significance and practical importance of the domestication process, while Chapter 6 discusses techniques used by farmers to obtain *D. rotundata* yams from yams collected in the wild.

The book concludes with various hypotheses to explain the phenotype transformations that take place as a result of domestication practices and their maintenance by vegetative propagation. Further studies are needed to assess these hypotheses, which are potential research topics for geneticists. Some are already being verified by combined teams of researchers from the North and South using the most recent molecular marker techniques.

Introduction

According to a study by IFPRI (Washington International Food Policy Research Institute; Scott *et al.*, 2000), sub-Saharan Africa accounts for nearly 96% of the world's yam production, while production in Africa increased by 183% between 1983 and 1996. Virtually all of this African output is confined to West Africa, with *Dioscorea rotundata* representing nearly 90% of all yams cropped in this region. The only exception is Côte d'Ivoire, where *D. alata* accounts for over 70% of all yam produced (Doumbia, 1998), even though 75% of the domestic trade involves *D. rotundata* yams (Touré *et al.*, 2003).

Yams were adapted to monocropping by societies belonging to what Miège (1952) called the 'civilization of the yam'. This adaptation occurred in savannah areas that had probably replaced a more wooded environment, as suggested by the presence of residual areas of mesophyll forest. Aubréville (in Schnell, 1971) put forward the idea that initially forested regions were converted to savannah as a result of human activities.

Scientists studying yam domestication were soon struck by the cultural importance of this crop. This topic has been discussed by several authors, including Coursey (1976), Seignobos (1992), Assogba (1993) and Allomasso (2001), who traced this trend back to the remote past of West African societies.

Societies belonging to the civilization of the yam are settled and well structured. *D. rotundata* can ensure a community's food needs throughout the year when all of its resources are tapped. Several West African ethnic groups have taken full advantage of these resources. For many reasons, others use only early-maturing cultivars, e.g. to bridge the gap between cereal harvests, because local climatic conditions are unfavorable for late-maturing cultivars, or because these late yams are scarce, low-yielding and thus unable to compete with *D. alata* yams (Côte d'Ivoire).

In 1939, Burkill was convinced that *D. rotundata* was the result of the domestication by African farmers of yams they found growing wild. However, this hypothesis was not scientifically confirmed until the end of the 20th century, when very powerful tools (enzymatic and molecular markers, flow cytometry) were used to reveal genetic relationships between *D. rotundata* and wild yams. Further insight was also acquired on traditional yam domestication methods. Firstly, on the basis of the findings of a survey of 150 farms in two regions of northern Benin (Dumont and Vernier, 1997a) and more piecemeal information obtained in other African countries. Relevant information was subsequently obtained in five in-depth studies conducted in Benin (Baco, 2000; Okry, 2000; Adoukonou, 2001; Allomasso, 2001; Mignouna and Dansi, 2002). Surveys in Nigeria (Vernier *et al.*, 2003) also indicated that yam domestication techniques used in several regions of the country were similar to those implemented in Benin. Lastly, Hildebrand (2003), in a study undertaken in southwestern Ethiopia, reported on a local form of domestication involving several wild yams with numerous similarities to the practices used in West Africa.

We felt that the time was now ripe to pool all available knowledge on the domestication of African yams from a substantial number of publications, unpublished and even unprocessed experimental results and, most importantly, field observations. Much of this information concerns Benin and Côte d'Ivoire but some was also collected in Guinea, Togo, Burkina Faso, Nigeria and Cameroon. The present review therefore covers most of West Africa in varying degrees, while extending into Central and East Africa on a number of occasions.

We venture beyond the scientifically proven results in our discussion and advance many hypotheses, some of which are based on very recent theories that have considerably broadened the scope of yam genetics. The future will judge the merits of the viewpoints proposed.

The technical terms are defined in a glossary at the end of the book.

The *Dioscorea rotundata* Poir. yam

Botanical aspects

There has long been considerable confusion regarding the yams *Dioscorea rotundata* Poir. and *D. cayenensis* Lam. In English-speaking West Africa, particularly Nigeria, they are known as 'white yam' and 'yellow yam', respectively, and pooled under the term 'Guinea yam'. Farmers in French-speaking Africa, on the other hand, do not make a clear distinction between *D. rotundata* and *D. cayenensis*, whereas a generic name is used for all other cultivated yams (*D. alata, D. bulbifera, D. dumetorum, D. esculenta*)—although the latter are not regarded as 'true yams' by many ethnic groups (figure 1). The diagnoses of Lamarck (1792) and Poiret (1813) proved to be too inaccurate to separate *D. cayenensis* and *D. rotundata* (in Miège and Lyonga, 1982). Finally, Miège still regarded *D. rotundata* as a subspecies of *D. cayenensis* in the 1968 edition of his *Flora of West Tropical Africa*. This botanical status, first assigned to *D. rotundata* by Grisebach in 1854, was endorsed by Prain and Burkill in 1919 and Chevalier in 1936 (in Coursey, 1976). Because of this confused situation, the concept of a *D. cayenensis-D. rotundata* species complex was proposed at the 1978 Seminar on Yams in Cameroon, funded by the IFS (International Foundation for Science, Stockholm, Sweden). This concept was then defended by Hamon (1987) as a way of "pooling all West African cultivated yams that are not bulbiferous and have entire leaves under the same name".

D. cayenensis and *D. rotundata* are yams domesticated from wild Dioscoreaceae of the Enantiophyllum Uline section that have speciated in Africa. They differ with respect to various botanical and genetic traits but have never been definitively separated. It is thus essential to present *D. cayenensis* before investigating *D. rotundata*.

D. cayenensis stricto sensu (Poiret definition) is found in West and Central Africa. In West Africa it coexists with *D. rotundata* but is not widely cropped, whereas virtually all yams cropped in Central Africa (mainly forested areas) are *D. cayenensis* and *D. alata*, but *D. rotundata* is generally not grown. *D. cayenensis* has numerous vernacular names because of its extremely wide geographical distribution range: Yaobadou for

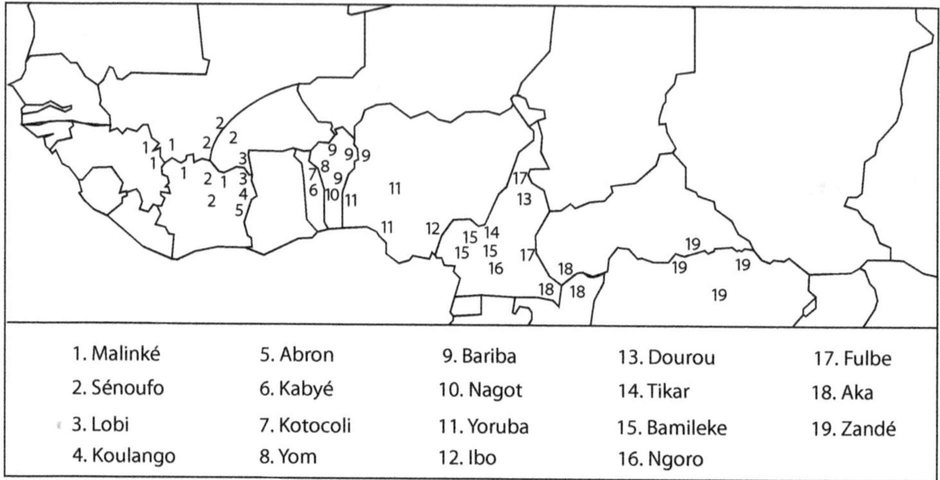

Figure 1. Geographical distribution of the ethnic groups mentioned in this book.

the Baoulé of Côte d'Ivoire (Hamon, 1987), Banoussé, Alakissa (Ikéni) and Kanlin for the Bariba, Nago and Adja peoples of Benin (Dansi *et al.*, 1999a), Ji oku and Ishu kpukpa for the Ibo and the Yoruba of Nigeria (Orkwor, personal communication), Mbip and Ekoto for the Dourou and the Bamileke of Cameroon (Dumont *et al.*, 1994; Mignouna *et al.*, 2002a) and Ako for the Teke-speaking peoples of Central Africa (N'Kounkou, 1993). This list is obviously far from exhaustive.

Several scientific studies and various observations have now been focused on the phyletic relations between *D. cayenensis* and other African yams of the Enantiophyllum section, including *D. rotundata*, but the situation remains unclear. It can be summarized as follows.

Terauchi *et al.* (1992), Ramser *et al.* (1997) and Chaïr *et al.* (2005) reported that *D. cayenensis* and *D. rotundata* bear the same chloroplast DNA (which would make them the same species), differing from that borne by *D. burkilliana*. Moreover, Terauchi *et al.* (1992) presented *D. cayenensis* as an interspecific hybrid on the basis of its nuclear ribosomal DNA characteristics. The female parent might be *D. rotundata, D. praehensilis* Benth, *D. liebrechtsiana* De Wild or *D. abyssinica* Hochst ex Kunth, which are all characterized by annual replacement of the vegetative organs and tuber. The male partner would be *D. burkilliana, D. minutiflora* Engl. or *D. smilacifolia* De Wild, which have a perennial base plate.

Some results of enzymatic or molecular marker analysis of total DNA point in the same direction as the ideas depicted above, while others diverge. Mignouna *et al.* (2002a) and Mignouna and Dansi (2002) distinguished between *D. cayenensis* and *D. rotundata* but did not divide them into separate species. Hamon (1987) suggested that *D. cayenensis* might be the product of interspecific hybridization but stressed the likely involvement of *D. burkilliana*. Other authors claimed that *D. cayenensis* is phyletically very close to or a domesticated form of *D. burkilliana* (Akoroda and Chheda, 1983; Onyilagha and Lowe, 1985;

Mignouna *et al.*, 1998; Dansi *et al.*, 2000b). Lastly, H.M. Burkill (1985) and Edeoga and Okoli (2001) considered *D. cayenensis* and *D. rotundata* to be two distinct species.

Several field observations support the potential phyletic proximity between *D. cayenensis* and *D. burkilliana*. Hamon (1987) reported that *D. burkilliana* produces tubers quite similar to those of *D. cayenensis* when it is cropped in Côte d'Ivoire. In the same country, *D. cayenensis*-type tubers were obtained from *D. burkilliana* seeds that had been cultivated in mounds for 3 years (Dumont, personal observation). The Bamileke of Cameroon still create *D. cayenensis* yams from *D. burkilliana* (Dumont *et al.*, 1994). Among the Nago-Idatcha peoples of central and western Benin, *D. cayenensis* is sometimes called Itschotinto, meaning 'millipede yam' (Dumont, personal observation), probably because the tubers of its wild ancestor resemble the large millipedes of the *Pachybolus* genus—but only *D. burkilliana* matches this description.

Various arguments linking *D. cayenensis* to *D. burkilliana* have been put forward. However, there is fairly high variability in the tubers although the vegetative organs are monomorphic. Excrescences on the epidermis (a trait identified by Burkill in 1918) are found in the Yaobadou yam of Côte d'Ivoire but not in the Mbip of northern Cameroon, while the tuber of the latter has a 'swan's neck' shape that has never been noted in Côte d'Ivoire. In addition, *D. cayenensis* is distinguished by two types of pre-tuber—large and irregularly shaped pre-tubers are common throughout West and Central Africa while the small spherical type is not as widespread. Many of the latter type have been observed in Cameroon, particularly in the Yaoundé region (Dumont, personal observation), and a few in central and western Benin (Dansi, personal observation). This morphological variability in the tuber suggests that *D. cayenensis* might have several origins, thus explaining the current divergence of opinion as to its botanical nature. If the pre-tuber is a relic of the fibrous base characteristic of wild yams with pluriannual or perennial vegetative organs, then it is quite likely that *D. cayenensis* actually has two ancestors. One would be *D. burkilliana*, whose elongated base grows plagiotropically and can exceed 50 cm, and the other *D. minutiflora*, which has a much smaller circular base. The latter has recently been regarded as a form of *D. burkilliana* (Mignouna and Dansi, 2003; Chaïr *et al.*, 2005), thus implying that the species is extremely polymorphic. However, these two yams still cannot be definitively regarded as the same species. Enzymatic markers have revealed two genetic groupings among *D. minutiflora* yams from Côte d'Ivoire (Hamon, 1987).

The octoploid nature (2n = 80, X = 10) of *D. cayenensis* has been established in several studies (Zoundjihèkpon *et al.*, 1990; Zoundjihèkpon, 1993; Hamon *et al.*, 1992; Dansi *et al.*, 2000b, 2000c). However, Dansi *et al.*, in a flow cytometry analysis, found three ploidy levels (4X, 6X, 8X) in *D. cayenensis* yams from Cameroon in 2001. This is the first time that *D. cayenensis* was shown to be a polyploid series, but this finding requires confirmation.

Only male *D. cayenensis* plants are currently known, and their fertility was found to be very low when IITA (International Institute of Tropical Agriculture, Nigeria) used them in controlled crosses (Dansi, 1999a). The species status attributed to *D. cayenensis* is therefore debatable. This yam seems to be the result of genetic accidents, notably ploidization, and—after collection by farmers—subsequent modification of their morphological traits through cultivation. It should therefore be regarded as a cultigen.

Enzymatic or molecular markers have revealed cultivated yams in West and Central Africa that belong to neither *D. cayenensis stricto sensu* nor *D. rotundata* but have genetic affinities with the former (Hamon, 1987; Zoundjihèkpon, 1993; Seniou, 1993; Dumont *et al.*, 1994; Dansi *et al.*, 2000b; Camara, 2001). These yams are male and appear—in the light of several cytogenetic studies (Zoundjihèkpon *et al.*, 1990; Zoundjihèkpon, 1993; Hamon *et al.*, 1992; Dansi *et al.*, 2000b, 2000c; Camara, 2001)—to be hexaploid ($2n = 60$, $X = 10$). Morphologically, they are extremely heterogenous. They can be divided into four groups on the basis of current knowledge:

– The Kangba yam of Côte d'Ivoire (Abron, Koulango, Agni, Baoulé and Djimini ethnic groups) and the Shoufing yam, its alter ego in the Bamileke region of Cameroon. They are primarily characterized by leaves with well separated lobes and a petiole armed with a thorn. In Côte d'Ivoire, Hamon (1987) identified several cultivars differentiated by the color of the tuber flesh: white, light yellow, dark yellow, violet, or yellow with mauvish areas.

– The Kpokpokpokpo yam (Agni ethnic group of Côte d'Ivoire). The tuber consists of a tight cluster of globular masses with a white or light yellow flesh. In a molecular marker analysis, Mignouna *et al.* (1998) showed that it shares 60% of its genome with *D. cayenensis*, but the other parent has not been identified. This is the first probable case of interspecific hybridization.

– An arbitrary grouping consisting of the Bolgo Nyu yam (central Burkina Faso) and the Kpeyou yam, its equivalent in the Kabye and Kotokoli regions of northern Togo; Makpawa and Ofegui yams in the Yom and Nago regions of Benin, respectively; and Bamba and Gban yams of the Malinke region of Guinea. Apart from the light yellow color of the tuber flesh, these yams are morphologically heterogenous. However, in general terms, their principal characteristics clearly distinguish them from the two previous botanical groups while partially linking them to *D. rotundata* yams, which are discussed later.

– The Baridjo cultivar of the Boko people of Benin. It has the leaves of *D. burkilliana* and the stem of *D. praehensilis* (Dansi *et al.*, 1999). Its enzymatic traits distinguish it from the Makpawa and Ofegui cultivars, which were classified in the previous group by Dansi *et al.* (2000b) and Mignouna and Dansi (2003). This is presumed to be a second case of interspecific hybridization. In Côte d'Ivoire, similar morphotypes appeared in the progeny of experimental hybridizations between the female cv Krenglé (*D. rotundata*) and *D. praehensilis* (Dumont, personal observation). The ploidy of this material was not measured.

Hamon (1987) linked the hexaploid cultivars of Côte d'Ivoire to *D. cayenensis*, while also suggesting that the latter might be an 'artifact' species deriving from two distinct wild species, with one giving rise to the Yaobadou yam (*D. cayenensis stricto sensu*) and the other to the Kangba yam (*D. cayenensis lato sensu*) and its associated hexaploid forms. The problem remains unresolved. There is an assumed phyletic relation between some or all of this plant material and wild yams with a perennial base plate, especially *D. burkilliana*.

The findings of chloroplast DNA analyses and enzymatic or molecular marker studies on total DNA generally differ. The reasons underlying this discrepancy require investigation and, moreover, the diversity of *D. cayenensis* and *D. burkilliana* yams must be explored in greater detail before examining the degree of mutual kinship and that between the two of them and with *D. rotundata*.

Geographical distribution

The African yam belt was described by Coursey (1967) as lying between western Cameroon and the Bandama River in Côte d'Ivoire, in the climatic area bounded by the 800 mm isohyet in the north and extending southwards to the Atlantic Ocean.

The area in which *D. rotundata* production is economically important does not fall exactly within these boundaries (figure 2). It starts in western Cameroon but extends as far as Upper Guinea, with a relatively empty space between the western border of Cameroon and Bandama River. It is highly concentrated in the savannah areas between latitudes 6° and 10° N.

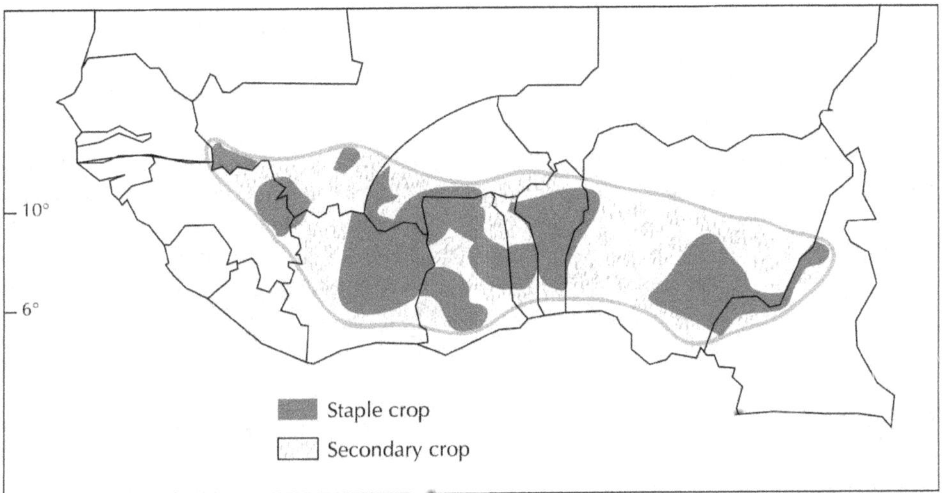

Figure 2. Geographical distribution of *Dioscorea rotundata* yams in West Africa.

The West African area in which *D. rotundata* is grown has changed over the last few decades. In southeastern Nigeria, commercial agriculture has pushed the yam cropping area northwards (Manyong *et al.*, 1996), while production has developed in Burkina Faso along the Ghanaian border in the vicinity of Ouagadougou, the country's capital and largest city. Urban market demand has also prompted *D. rotundata* production to southern Benin. The same phenomenon is apparent in the extreme southern part of Mali, in the Segou region (near Bamako), and in the area bordering on Senegal, which supplies the Dakar market.

Yams are also reported to be commercially cropped in savannah areas of the Central African Republic (Dumont *et al.*, 1994) and southern Chad (Mbailao Kemdigao, 1998). Few inventories have been made of the plant material used, so the proportion of *D. rotundata* yams cropped is unknown.

Botanical origins of *Dioscorea rotundata*

The phyletic relations between *D. rotundata* and wild yams have long been the focus of scientific investigation. Chevalier linked the Soussou cultivar of northern Benin to *D. praehensilis* in 1920 and to *D. lecardii* De Wild in 1936. In 1920, he classified cv Sopere of Côte d'Ivoire as a *D. praehensilis* yam. Burkill (1939) believed that *D. rotundata* derived from *D. abyssinica* or from another wild yam of the same type (possibly *D. lecardii*). In his 1952 thesis, Miège claimed that *D. abyssinica, D. sagittifolia* Pax, *D. praehensilis, D. liebrechtsiana* and *D. mangenotiana* J. Miège are possible parents of *D. rotundata*, whereas Coursey (1976) suggested only *D. praehensilis*. The relevance of these different opinions will become apparent over the course of this book.

D. rotundata has been linked to the wild yams *D. praehensilis* and *D. abyssinica* in several recent scientific works (Hamon, 1987; Terauchi *et al.*, 1992; Zoundjihèkpon, 1993; Dansi, 1995; Dansi *et al.*, 1999; Scarcelli, 2002). The latter two authors showed that the geographical distribution of the different forms of *D. rotundata* in Benin reflected that of *D. praehensilis* and *D. abyssinica*. Both of these wild species can give rise to *D. rotundata*, and all three are tetraploid (2n = 40, X = 10) according to Zoundjihèkpon *et al.* (1990), Hamon *et al.* (1992), Zoundjihèkpon (1993), Gamiette *et al.* (1999), Dansi *et al.* (2000b) and Camara (2001).

D. rotundata, the focal point of our study, can be defined as a group of cultivated Dioscoreaceae yams of African origin belonging to the botanical section Enantiophyllum Uline, with a short annual vegetative cycle (6–8 months), a tuber with a long dormancy period (3–5 months) and a slightly or non-pigmented (creamy or white) flesh. We thus make a distinction between *D. rotundata* and *D. cayenensis*. These two yams probably have a common ancestor at a yet undetermined phylogenetic level, but only the former is involved in domestication in West Africa.

Organization of diversity in *Dioscorea rotundata*

Yam diversity on African farms

D. rotundata is subdivided into varieties or cultivars. Both terms are used in the scientific literature on cultivated yams, although they have slightly different meanings. When a cultivar is mentioned hereafter, the abbreviated form 'cv' (plural 'cvs') will be used.

For the African farmer, a yam cv is identified by its common name, which often contains technical or historical information. Yam cvs are best differentiated on the basis of their tuber traits. The characteristics of the vegetative organs are sometimes, but not always, also distinctive markers. Farmers, on the other hand, consider yam cvs as sets of technical criteria consisting of agronomic requirements, harvesting time, cooking quality and storage life.

It is not known how many *D. rotundata* cvs are cropped in Africa. According to Martin and Sadik (1977), the *D. cayenensis-D. rotundata* complex includes between 500 and 2,500 West African cvs. 166 names have been recorded for *D. rotundata* yams in Côte d'Ivoire and 311 in Benin (Hamon, 1987; Dansi *et al.*, 1999), but many of these names are probably synonymous since a large number of ethnic groups use this vegetatively propagated plant material. Regional surveys clearly revealed the range of yam plant material used on farms in Benin. Each farmer grows about 30 *D. rotundata* cvs at most, and around 10 are generally cropped in over 70% of the yam-growing area (Dumont, 1997; Baco, 2000; Okry, 2000; Vernier and Dossou, 2002).

In eastern and central Côte d'Ivoire, Hamon (1987) noted only 5 to 15 *D. rotundata* cvs per village and even fewer in the western region. In Upper Guinea, the main yam-growing region of Guinea, very few *D. rotundata* cvs are cropped and several of these are shared with neighboring Côte d'Ivoire (Dumont, 1993).

The fact that yams were only quite recently domesticated in these countries could account for the narrow range of cvs grown there. The French explorer René Caillié (1932), who traveled from Kankan (Guinea) to Tengrela (Côte d'Ivoire) between June 1827 and January 1828, conveyed a substantial amount of information about the food habits of the people he encountered, but he did not mention yams in his description of his journey to Timbuktou. Based on the collective memory of the Malinke, Camara (2001) estimated that yam domestication had been practiced in Guinea for 200 years. In reference to a historical benchmark, Sekouba (2001) indicated that *D. rotundata* yams were known in the region before 1840. Lastly, the Abron farmers of Côte d'Ivoire (Dumont, personal observation) say that local domestication leading to *D. rotundata* yams dates back to the arrival of the Akan from Ghana, i.e. in the 17th or 18th century (Assogba, 1993 and Rougerie, 1982, respectively). In Côte d'Ivoire, the entire sequence of intermediate forms between *D. praehensilis* and *D. rotundata* are still being used, which further confirms the recent nature of domestication in this country. This trend is also noted in southern Benin, where *D. praehensilis* has been recently domesticated and Gnidou-type *D. rotundata* yams created to supply the attractive nearby Cotonou market.

The *D. rotundata* species is divided into two groups, double-harvest (or early-maturing) yams and single-harvest (or late-maturing) yams. Many late cvs have been grouped together under the Nigerian-Beninese name kokoro yam (Dansi *et al*, 1999). However, kokoros are generally *D. rotundata* yams that are processed into dried chips, which excludes late yams from Côte d'Ivoire and Guinea.

Double-harvest cvs give two crops during the same growing season, each producing a different type of tuber. The plants initially produce one or two long (0.25–0.75 m), heavy (2–10 kg) tubers and these are harvested or 'milked' 3 to 5 months after emergence, with care taken to not damage the root system. These early tubers have a short storage life and are used for consumption. After removal of the first tubers, each plant usually produces a large number of new tubers of highly variable size that are more or less joined at the head. The overall weight of these second-growth tubers ranges from 0.5 to 8 kg. They are harvested several weeks after the end of the annual vegetative cycle, i.e. after the aerial parts of the plant have died back. A 4-year study covering most of the savannah area of Côte d'Ivoire and based on a sample of 4,340 plants reported an average second-harvest yield of 6.7 t/ha in traditional agriculture conditions, as compared to a first harvest-yield of 11 t/ha (Dumont and Kouakou, in press). In Benin, 754 pairwise measurements over 3 agricultural years showed an average second-harvest yield of 9.3 t/ha and a first harvest-yield of 14.6 t/ha (Vernier and Dossou, 2000). Most second-harvest tubers are very fibrous and therefore unsuitable for consumption. However, they have a long storage life if they are regularly desprouted following the dormancy period. They are used as seed tubers (sometimes cut into setts) for vegetative reproduction—the usual method of propagating *D. rotundata* yams in agriculture. This double-harvesting technique differs markedly from the harvest series sometimes used with *D. dumetorum* yams, as the latter simultaneously produce several tubers that have different growth rates.

Single-harvest cvs give one crop per year, at the end of the vegetative cycle. Each plant produces several tubers, with the number and individual weight of these tubers varying considerably according to the cv, environmental conditions and cultivation techniques. As these tubers are short, they can be left in mounds in the field until the middle of the dry season without increasing the risk of damage during harvesting. Losses due to rotting during subsequent storage are therefore low. Tubers of single-harvest cvs can be stored for 3 to 6 months after harvesting. The harvested crop is divided into two parts: seed tubers to be planted the following crop year and tubers for consumption.

The criteria currently used to select single-harvest *D. rotundata* cvs in some parts of West Africa contrast with those used previously. In the past, farmers primarily cropped cvs that produce a small number of tubers since they wanted to obtain larger individual tubers, with a consequent reduction in peeling losses (Dumont, 1971). Over the last 25 years, however, selection pressure has been in the opposite direction in Benin, Togo and the Yoruba region of Nigeria, where urbanization has created a commercial demand for dried yam chips (Vernier *et al.*, 1999). In this new socioeconomic context, and probably also due to the introduction of recent domestic cvs (Mignouna and Dansi, 2002), there has been a sharp rise in demand for kokoro yams, which were previously not used to a great extent in agriculture (from Côte d'Ivoire to Nigeria). Kokoro yams give four to ten tubers per plant, each weighing 200 to 800 g. This naturally occurring form of tuberization is morphologically and physiologically very similar to that triggered artificially to create seed tubers from double-harvest cvs.

Double-harvest yams can also be harvested once rather than twice a year. This is necessary when late tuberization makes early harvesting inappropriate. It is also now a commercial strategy that has been deliberately developed in Côte d'Ivoire to prolong the supply of fresh tubers to urban markets, while limiting post-harvest losses (Dumont and Kouakou, in press). On the other hand, double-harvesting is pointless with late yams, as their numerous tubers do not grow to a sufficiently large size for cooking or marketing early in the growing season.

Other *D. rotundata* yams fall into an intermediate category. Whether they are harvested once or twice a year depends on the climatic conditions, variations in the cropping system or production strategy. The Krengle cv, for instance, is generally a single-harvest yam in Côte d'Ivoire but it is widely harvested twice a year in north-eastern Guinea, an area with more regular rainfall. Each plant generally produces two or three different-sized tubers, the largest of which can weigh up to 5 kg. The medium-sized tubers (± 1 kg) are used as seed tubers. Cvs of this type are hereafter referred to as mixed cvs.

Surveys conducted among the Bariba of Benin (Dumont, 1997; Vernier and Dossou, 2000) show that farmers there divide *D. rotundata* yams into two groups, i.e. single-harvest and mixed cvs (Yassounou yams) and double-harvest cvs (Tamdwe or Tamdoua yams). The second group is subdivided (sometimes unclearly) into two agronomic types. There may be another explanation for this division. In the local language, Tamdwe means 'male yam', but the term has nothing to do with the sexual characteristics of these plants. It recalls the observation of Seignobos (1998) that the Dourou of Cameroon regarded *D. alata* yams as females because of their historically recent introduction into local agriculture. Transposed to the Bariba region of Benin, this symbolic association implies that Yassounou yams appeared later than Tamdwe yams. This idea recurs throughout our study.

State of the scientific knowledge

Morphological and physiological diversity

The division of *D. rotundata* yams into double-harvest and single-harvest cvs has long been recognized as an agronomic trait, but this trait has only very gradually been associated with the morphological and physiological diversity of the plant material.

An early study of 56 *D. rotundata* yams collected in northern Benin showed that a rough dividing line could be established between single- and double-harvest *D. rotundata* yams on the basis of their morphological and physiological traits (Dumont, 1977).

When analyzed on the basis of an ascending hierarchical classification (Miège, 1979), the data of the above study revealed two sets of cvs, corresponding to single- and double-harvest yams. There was also some evidence on the existence of several intermediate cvs combining the traits of both forms. These usually corresponded to what we have termed mixed cvs.

The findings of Dansi *et al.* (1999) separated *D. rotundata* yams of Benin more precisely. All double-harvest cvs could be morphologically linked to a single wild ancestor, i.e. in some cases *D. abyssinica* and otherwise *D. praehensilis*. In contrast, single-harvest cvs appeared to combine the morphological traits of both wild parents, despite several exceptions (e.g. the number and shape of tubers produced by kokoro yams). The authors therefore regarded them as interspecific hybrids. The same study also demonstrated the existence of intermediate cvs combining the traits of single- and double-harvest yams, thus confirming the observations of Miège (1979).

Mixed and single-harvest cvs have traits that do not occur in double-harvest cvs. Each plant regularly produces two or more tubers and these have a much longer storage life than tubers of double-harvest cvs. Coalesced thorns are sometimes observed on the stems, usually accompanied by punctiform asperities in patches or elongated areas on the basal part of the stem.

Other traits appear to be more cv-specific. The fasciation of young stems and 'boulage' phenomena (premature tuber formation) have so far only been observed in Boni Oure, a mixed cv from northern Benin. Occasional bulbil production has been noted in Krengle and Gnan yams from Côte d'Ivoire (Dumont, personal observation) and in *D. praehensilis* yams in the course of domestication (Hamon *et al.*, 1995; Vernier, personal observation). Bulbils are usually produced when *D. rotundata* yams are cultivated *in vitro*. Bulbil production probably always reflects a physiological disturbance. Several traits appear to be specific to single-harvest yams: a pear-shaped, bifurcated tuber (numerous cases in Benin, Togo and Nigeria, and cv Kroukroupa of Côte d'Ivoire), attenuation of the voluble character of the stem (several kokoro cvs) and substantial dwarfing of the vegetative organs, e.g. in cv Tam Saan (literally 'groundnut yam') of northern Benin. This latter phenomenon might be due to genetic mutations and/or consanguinity. In Ghana, *D. rotundata* seeds irradiated with cobalt 60 produced dwarf plants (Klu,

1993), while a dwarf *D. rotundata* was obtained in India from the seed of a half-sib cross performed by IITA in Nigeria (Nair and Abraham, 1992).

Unusual variations occasionally affect the biological functioning of *D. rotundata* yams. The Lansome cv from the Dourou region of Cameroon was found to produce very spiny stems followed by totally spineless ones on the same plant (Dumont, personal observation). With several Beninese cvs of the Boni Oure group, trilobed leaves appear at the onset of the vegetative cycle (Dumont, personal observation) whereas this trait has only been documented in two Malagasy species, i.e. *D. proteiformis* H. Perr. and *D. flandra* H. Perr. (Burkill and Perrier de la Bathie, 1950). Dansi (1999) reported that seed tubers of the Soagona, Nonforwou and Mondji cvs from Benin sometimes develop biologically on a pluriannual basis instead of breaking up, a trait that is associated with the wild species *D. mangenotiana* J. Miège. In short, the morphological or physiological expression of the genome can vary during the vegetative cycle or between vegetative cycles, which may reflect an unstable state. It will be seen later that this situation is an advantage for domestication but there is a potential risk of phenotype destabilization for the cultivated plant material.

The stems or primary branches of *D. rotundata* yams, like those of *D. abyssinica*, sometimes fuse together over part of their length. This self-grafting is relatively surprising in plants that are usually classified as monocotyledons. However, various scientific works have shown that some morphological and anatomical traits of Dioscoreaceae yams are typical of dicotyledons (Miège, 1952; Lawton and Lawton, 1967; Ayensu, 1972; Degras *et al.*, 1977; Degras, 1986). Conlan *et al.* (1995) even suggested that Dioscoreaceae might be closer to dicots than to monocots. This was based on the results of an analysis of the DNA of codons that initiate the formation of the principal proteins stored by the tuber. From a practical standpoint, these ideas could pave the way to substantial and original research, provided that grafts between different genotypes can be successfully achieved. Work on this subject has already been undertaken in Nigeria using two *D. alata* cvs, with promising initial results (Shiwachi *et al.*, 2003).

Some double-harvest *D. rotundata* yams occasionally put out roots and tubers at the nodes of stems in contact with the soil. This phenomenon was observed by Dansi *et al.* (1999) in the Agogo, Douroubayesirou, Gnidou, Nonforwou, Ourtchoua and Soagona cvs of Benin and has also been reported in cv Wakourouni from the Malinke region of Guinea (Dumont, personal observation). In this region and the Bariba region of Benin, such an event is regarded as a manifestation of occult powers heralding a death in the grower's family. This cultural convergence probably highlights an ancient Malinke migration to Benin, evidence of which can be found today in the patronymic clan names of the Bariba (Bio Bigou, 1994). Botanically, the rooting of stems links *D. rotundata* to three wild species of the Enantiophyllum section, i.e. *D. togoensis* Knuth, *D. minutiflora* and *D. smilacifolia*, and also to the *D. hirtiflora* Benth species of the Asterotricha section. Another link will be established between these two botanical sections later. This double example indicates the wealth of information that can be obtained from African yams.

D. rotundata and *D. cayenensis* yams have been classified according to their morphological and physiological traits in Côte d'Ivoire (Hamon, 1987), Togo (Kassamada, 1992; Seniou, 1993) and Benin (Dansi, 1995; Dansi *et al.*, 1999). The same procedure was followed in all three countries: using IPGRI (International Plant Genetic Resources Institute, Italy) descriptors, the plant material was divided into cv groups that had a large number of common morphophysiological traits. There were 13 groups of *D. rotundata* cvs in Côte d'Ivoire, as compared to 23 groups in Togo and Benin. Some groups were common to two or three countries (Dansi, 1995; Dansi *et al.*, 1999). Multiple-tuber *D. rotundata* yams (kokoro yams *stricto sensu*) were well represented in Benin, Nigeria and Togo, but none were found in Côte d'Ivoire. In this country, only the tuber of cv Kroukroupa had similarities with cv Kagourou of Benin, which Dansi (1995) classified as a kokoro yam. The fact that domestication is a recent phenomenon should be taken into account when assessing the limited diversity of this type of yam in Côte d'Ivoire.

Genetic diversity

Information from enzymatic marker analyses

The enzymatic polymorphism of *D. rotundata* and *D. cayenensis* yams has been studied in groups formed on the basis of morphophysiological criteria using five or six enzymatic markers. The same electrophoresis technique was used in Côte d'Ivoire and Benin but not in Togo, so only the results obtained in the first two countries can be compared.

These studies showed that the enzymatic diversity of *D. rotundata* was far wider than its morphological diversity, with little concordance between the two. Moreover, the enzymatic markers were unable to distinguish between early, mixed and late-maturing *D. rotundata* cvs in Côte d'Ivoire (Hamon, 1987) or Benin (Dansi *et al.*, 2000b).

Five cv groups in Côte d'Ivoire appeared to be monoclonal, but in three cases the sample sizes were small. The results indicated that the other eight groups were probably polyclonal. Figure 3 shows the genetic structure of *D. rotundata* yams from Côte d'Ivoire. The Sopere group is presented as the least sophisticated domestication stage, as supported by two arguments. Hamon (1987) reported that only the Sopere group is linked to wild Bayere yams (*D. praehensilis* ecotypes) via Abron and Koulango populations in the eastern part of the country. The Sopere group is also the only one to include cvs of inferior cooking quality (Kpassadjo and Sanata).

In the UPGMA dendrogram constructed in Benin (Dansi *et al.*, 2000b), the 80% similarity threshold separates 13 of the 23 cv groups established for the country. The 10 remaining groups are scattered among the previous groups. Five cv groups might be monoclonal but this requires confirmation since a very small sample was analyzed in three cases. The authors provide 86 paired observations that establish a parallel between the size of the analysis sample and the number of genotypes detected. The statistical analysis revealed a very close correlation (r = + 0.833 for n-2 = 84) between the two variables. In other words, increasing the size of each set studied (cv group, cv, morphotype) should reveal greater genetic diversity, possibly including some cvs that are currently regarded as monoclonal.

Overall, the *D. rotundata* yams of Côte d'Ivoire and Benin appear to have a similar genetic structure. In each country, there are indications of ongoing continuity encouraged by repeated domestications of a wild plant material and further complicated by kin relationships, thus limiting genetic diversity.

Moreover, *D. rotundata* has been shown to have low genetic diversity in two different respects. Zoundjihèkpon (1994) confirmed the origin of a progeny obtained in Côte d'Ivoire by controlled hybridization of Zrezou (male) and C.20 cvs, with the former resulting from local domestication and the latter originating from eastern Nigeria

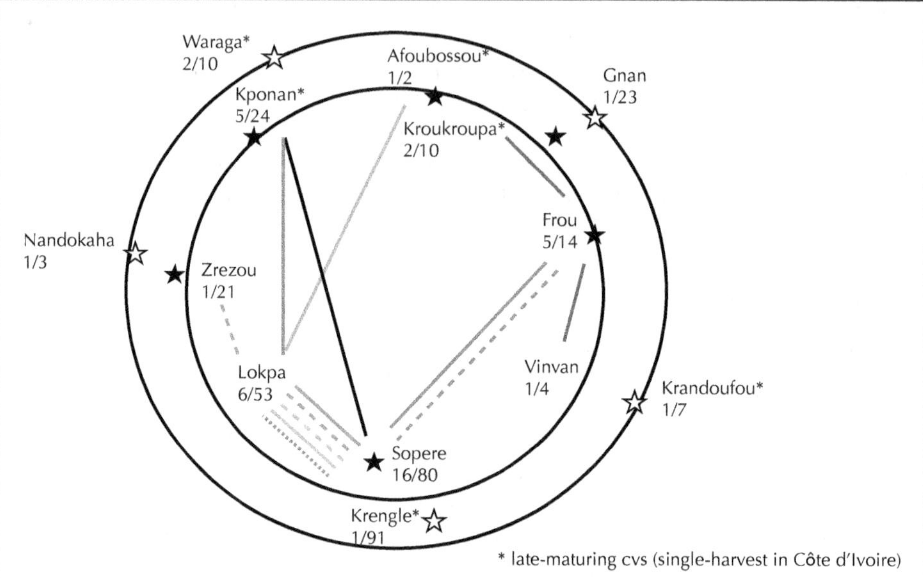

Explanations and comments

The name of each cv group is accompanied by two numbers. The first indicates the number of zymotypes revealed by electrophoresis, the second the size of the sample.

Each of the five groups in the outer circle includes specific genotypes and is therefore genetically isolated within the limits of the enzymatic markers used in the study and of the analysis sample.

The inner circle includes eight cv groups, representing a total of 37 zymotypes. Seven of these are common to two or more groups. Each genetic bridge is indicated by a line of a different colour and type.

The Sopere and Lopka groups are the most strongly interconnected, with five zymotypes in common, and they are also linked to the other groups on the inner circle. They might therfore constitute an important domain of domestication in Côte d'Ivoire. The selection order would be from Sopere to Lopka. The analysis shows that 32.5% (16/80) of the Sopere would therefore be the primary origin of *D. rotundata* yams in Côte d'Ivoire. In addition, it is regarded as a domesticated form of *D. praehensilis* by Chevalier (1920), H.M. Burkill (1985) and Hamon (1987), meaning that all the cv groups enzymatically linked to Sopere would be related to *D. praehensilis*.

The same relationship might also apply to four of the groups on the outer circle. Hamon (1987) regarded Krengle as a domesticated form of *D. praehensilis*, i.e. a wild species that is probably also the origin of Krandoufou, Gnan and Nandokaka cvs, all of which have spiny roots. Moreover, the first two are widely established in the preforest ecosystem where the wild species *D. abyssinica* is absent.

Figure 3. Genetic structure of *D. rotundata* yams from Côte d'Ivoire. Graphic representation based on the enzymatic characterizations of 13 cv groups conducted by P. Hamon (1987).

(Dumont *et al.*, 1994). The genetic compatibility of these *D. rotundata* yams thus extended over a vast geographical area, but it is not known to what extent the idea can be generalized. She also observed 12–50% heterozygosity among the progeny of several controlled hybridizations between double-harvest *D. rotundata* cvs. Allelic diversity is thus not necessarily wide in these yams, even if they are dioecious. The various possible explanations for this situation are discussed in detail later.

Enzymatic marker analyses do give an indication of the genetic organization of *D. rotundata* yams, but this tool did not always effectively discriminate between yams in the different studies. There are two main problems.

Firstly, dividing the plant material into morphological groups easily leads to error, as it is often difficult to establish the boundaries of these groups for *D. rotundata* yams because their variability overlaps to varying degrees. This would largely account for the divergence generally observed between their morphological and enzymatic diversity. This concept was successfully demonstrated by Mignouna *et al.* (2002b), who established a close agreement between morphotypes and zymotypes among six groups of morphologically very distinct cvs from Cameroon.

Secondly, the combination of enzymatic markers used is not powerful enough to separate certain genotypes. Hamon (1987) presented the Krengle cv of Côte d'Ivoire as monoclonal on the basis of 91 analyzed plants, whereas the Senoufo and Malinke farmers of the Dikodougou region divide it into 19 units on the basis of tuber shape and agronomic and technical traits such as the period of production, production potential, drought tolerance, peeled yield and cooking quality (Tokpa and Dumont, 1995). It is hard to accept that a single genome could be responsible for the wide diversity of reported traits. Dansi *et al.* (2000b) noted a similar situation with the morphologically homogenous cv Kinkerekou of northern Benin. Their analysis of 25 plants revealed two zymotypes, whereas local farmers consider that this cv has far greater diversity, particularly at the biological level. Here again, enzymatic markers seem to be insufficiently discriminating.

Information from molecular marker analyses

Terauchi *et al.* (1992) used the RFLP (restriction fragment length polymorphism) technique to analyze chloroplast DNA and ribosomal DNA but failed to separate *D. rotundata* from its putative wild parents and no hybrid forms were detected. Similar results were obtained by Chaïr *et al.* (2005) with chloroplast DNA.

Mignouna *et al.* (1998) applied the same technique to assess nuclear DNA and found that cv Bakokae of northern Cameroon, cv Noworfou of Nigeria, Gnidou and Terkokonou cvs of Benin and cv Zrezou of Côte d'Ivoire were genetically closely related. On the other hand, some other cvs found in Benin or Côte d'Ivoire were not found in Cameroon, but the Cameroon sample included only a small proportion of the *D. rotundata* yams cultivated there.

The AFLP (amplified fragment length polymorphism) technique was used by Camara (2001) to analyze total DNA from four Guinean double-harvest *D. rotundata* cvs in parallel with seven local *D. abyssinica* yams. The cvs appeared to be polyclonal to various extents and a neighbor-joining dendrogram separated them into two groups unrelated to *D. abyssinica*. The genetic characteristics of seeds of one of the cvs studied were found to differ markedly from those of the maternal parent but resembled those of Guinean *D. abyssinica* yams.

The same method was used by Scarcelli (2002) in Benin to compare *D. praehensilis*, *D. abyssinica* and *D. rotundata* yams that had been domesticated in the past or were in the course of domestication. The two wild species were divided into two categories according to whether they had been collected close to or remote from a *D. rotundata* cropping area. The scientific value of this pioneer study has been recognized and rewarded. The following is a summary of the main findings (with a comment on each):

– The three species correspond to distinct groupings with a common genetic background. One of the observations of Camara (2001) indicates that individuals with 50% of their genome in common can appear to be genetically very distant from one another. Large genetic differences thus do not necessarily indicate a distant genetic relationship.

– *D. rotundata* yams in the course of domestication appear to be genetically very close to wild yams or to long domesticated *D. rotundata* cvs. The latter situation predominates in northern Benin, where there is a long history of yam domestication. In this case, plants genetically very close to cultivated yams would have become established in the wild.

– *D. rotundata* yams of northern Benin (56% of the total analyzed from Benin) can be divided into seven groups. One group contains traditional late cvs of the Bariba region. The others consist of early or mixed cvs. Four of the seven groups are genetically close to yams in the course of domestication and therefore linked to wild yams that appear to have incorporated genes from cultivated yams. This is not the case with the other three groups, including single-harvest *D. rotundata* yams, and could indicate a loss of sexual function, sterility or recent introduction into local agriculture.

– With five cvs (Worou binsi, Tabane, Douroubayesirou, Ahimon, Boni Oure), two or three different genotypes were found in the two to four plants studied per cv. There is a larger or smaller genetic distance between genotypes of the same cv. The polyclonal structure of the cvs, already indicated on several occasions, is again apparent. A much broader sample should be assessed to obtain reliable results in a genetic inventory of such plant material, and also with respect to any derived information on phyletic links with wild yams or yams in the course of domestication.

– northern Benin, whether *D. abyssinica* and *D. praehensilis* yams are collected close to or remote from *D. rotundata* crops does not markedly alter the dendrogram distribution of *D. rotundata* yams in the course of domestication. Shifting agriculture has been a feature of the region since ancient times and few wild yams have escaped the associated gene flows. One exception might be the Upper Oueme protected forest, which was inhospitable until the late 1970s because of the high onchocerciasis risk. No samples from this environment were included in the study.

The RAPD (random amplified polymorphic DNA) markers used by Dansi *et al.* (2000a) on the DNA of single-harvest *D. rotundata* yams from Benin (including, according to farmers, kokoro yams recently introduced from Nigeria) revealed wide genetic diversity spanning four groups. Virtually all traditional cvs from northern Benin belong to the same group. This separates them from recently introduced kokoro cvs and confirms—as traditionally considered—that they were domesticated locally.

In short, molecular markers have a high discriminating power. In two different studies, they showed that late-maturing *D. rotundata* yams had distinctive genetic traits, thus separating them from the other yams, while still having some diversity. This finding suggests that this type of yam is of hybrid origin, although other hypotheses are not excluded.

Cytogenetic features of *Dioscorea rotundata* yams and their wild relatives

At present, *D. rotundata* yams are widely regarded as being tetraploid, but this preconception could be challenged in the future. Daïnou *et al.* (2002) have summarized experimental results which suggest that *D. rotundata* might be diploid:

– When monitoring two enzymatic traits in the progeny of a monoecious clone of cv Gnidou from Benin, the authors noted segregations corresponding to diploidy ($2n = 40$, $X = 20$).

– Zoundjihèkpon (1993) pointed out that disjunctions observed in the three enzymatic systems of the progeny of a controlled cross between Zrezou (male) and Sopere (female) cvs from Côte d'Ivoire could match the theoretical 1-2-1 segregations expected in the case of diploid individuals and a cross between heterozygotic partners. Supposedly tetraploid yams thus behaved in a diploid manner. Furthermore, Mignouna *et al.* (2002b) concluded that the *D. rotundata* genome is allopolyploid ($2n = 4$, $X = 40$).

– In a study of enzymatic polymorphism in a *D. praehensilis* population in Côte d'Ivoire, Hamon and Tio-Touré (1982) indicated that the numbers observed in this population corresponded to the theoretical numbers of a panmictic diploid population in a system controlled by a gene coding for an active protein in homo- or heterodimer form. According to Daïnou *et al.* (2002), this situation is confirmed by the results of a yet unpublished Beninese study.

Chemical diversity

Many *D. rotundata* yams are more or less bitter. The bitterness is usually greater at the tip (or distal extremity) of the tuber, which is why this part is often removed before pounding. Physiological immaturity of the tuber (the stage preceding senescence of the vegetative organs) and unfavorable environmental conditions (drought, waterlogging) are factors that increase the bitterness. The latter is due to the presence of saponins and tannins, which both occur in the tuber flesh in the form of several chemical compounds (Osagie, 1992.)

Saponins mainly contain sapogenins, which are known to have high efficacy in medical applications (Degras, 1986). Domestication appears to have eliminated this metabolite from *D. rotundata* yams since they play a minor role in the traditional pharmacopoeia (Dumont, personal observation). Sapogenins, however, are present in *D. abyssinica* (Martin, 1969) and probably also in *D. praehensilis*, which are regarded as the ancestors of *D. rotundata*. It is noteworthy that both of these wild yams are used to treat various abdominal pains in Guinea (Dumont, 1993), Central Africa (N'Kounkou, 1993; Dounias, 1996) and Benin (Baco, 2000; Allomasso, 2001).

Tannins are polyphenols (Osagie, 1992) and thus the main compounds responsible for the bitterness of *D. rotundata* yams. Relatively bitter cvs can be found amongst both early and late yams. In the fairly recent past, when food self-sufficiency was a priority strategy, yam bitterness was probably promoted as a security factor, providing a certain degree of protection against predators (Osagie, 1992). There is currently little commercial demand for bitter *D. rotundata* yams. The one exception is cv Krengle of Côte d'Ivoire, but its slight bitterness is accompanied by a pleasant chestnut-like taste. This combination of flavors is highly appreciated by local people, which explains why cv Krengle is one of the most widely cultivated *D. rotundata* yams in the country.

Again according to Osagie (1992), polyphenol oxidation (when the tuber is cut) is responsible for browning of the flesh. In traditional cuisine, this deterioration in quality seems to be specific to cvs with a high mucilage content (Dumont, personal observation). The same alteration occurs in food prepared from dried yam chips. What starts as a pinkish-white flour ends up as a light- or dark-brown dough (amala) whose taste is modified to various extents by lactic fermentation. The intensity of these phenomena depends on the cv used, the length of soaking after parboiling (Akissoé *et al.*, 2003) and the drying rate (Kayodé *et al.*, 2005).

A number of chemical traits noted in Beninese *D. rotundata* yams can be tentatively mentioned. The skin of cv Soussou ('bee' in the vernacular) which is cropped by the Bariba peoples has an irritant effect when harvested, probably because of its high raphide content. The Bariba cv Gbera is galactogenic and therefore reserved for women with breast-feeding problems. The active metabolite in the latter yam has not yet been identified (Dumont, 1997). More generally, surveys have revealed that yam chips have therapeutic properties that can alleviate three different disorders, i.e. hyperglycemia, diabetes and hemorrhoids (Dumont, 1995). *D. dumetorum* is also known to have somewhat similar therapeutic properties, likely due to the alkaloid dioscoretin it contains (Iwu *et al.*, 1990). A patent has been filed to extract an antidiabetic medicine from this yam (Genetic Resource Action International, 2002). Note that the therapeutic properties attributed to dried chips produced from kokoro cvs have not been reported for dishes prepared using fresh kokoro tubers. Processing thus seems to be responsible for, or enhance, the medicinal effect of the chips.

Cooking quality

Fresh tubers are used to prepare traditional food dishes. From Côte d'Ivoire to Nigeria, yams are most appreciated eaten in pounded form (so-called futu or fufu) a dough obtained by mortar pounding tubers that have been boiled until soft. These countries account for over 90% of *D. rotundata* production in Africa (according to FAOSTAT, 2003). North of this area and throughout Central Africa, yam preparation is simpler—the tubers are just boiled and served with various stews. Recent studies have shown that most inhabitants of urban areas of West Africa eat boiled or fried yams, often as a snack away from home (Bricas *et al.*, 2003). The use of yam flour (produced by milling dried chips) is another emerging habit. Yam flour is very well adapted to urban cooking requirements and is used to prepare (5–6 min) a dough called amala, a staple or occasional food for about 50% of the population of Cotonou (Benin) and towns of southwestern Nigeria. Amala is not perceived as a substitute for futu but rather as a food in its own right (Bricas *et al.*, 1997).

Two quality criteria apply to food prepared from fresh tubers, i.e. the product must have a pleasant flavor and an attractive appearance. It thus should not become bitter or acquire an unpleasant color and texture during cooking. Most old domesticated cvs that are marketed commercially meet these criteria, but the situation is much more varied with cvs from areas where domestication is a relatively recent phenomenon (e.g. Gnidou yams from southern Benin).

Culinary requirements vary depending on whether the fresh tubers are just boiled or converted into futu. When eaten boiled, the flesh must remain firm and structurally homogeneous after cooking. Pounded yam must be lump-free, elastic and firm enough to form a ball with the fingers. Ease of pounding is another characteristic that the women take into account.

Yam cvs that produce good quality pounded yam are usually also appreciated boiled. However, the quality of the pounded product obtained with some cvs is poor and these yams are thus only consumed in boiled, fried or braised form. As already mentioned, all *D. rotundata* yams in Côte d'Ivoire that are unsuitable for futu-making belong to the Sopere group, which probably represents the initial stage in the local domestication of *D. praehensilis* forest yams. A similar situation exists in southern Benin with Gnidou yams. These have retained strong affinities with *D. praehensilis*, their wild ancestor, and are almost exclusively eaten boiled as they are considered to be of poor cooking quality (in the vernacular, Gnidou means 'yam appreciated only by cattle'). Yams unsuitable for futu-making would thus correspond to the lowest stage of domestication and be closely related to the wild forest parent *D. praehensilis.*

Kokoro cvs generally produce a pounded yam of excellent quality. This is a first indication that they might originate from wild plant material that has long been manipulated by man or already cropped.

Cooking quality varies markedly at each domestication level. Differences between cvs involve their degree of organoleptic quality, as indicated above, but also criteria such as optimum harvesting time or storage life, fiber content and peeling loss.

Apart from problems relating to bitterness and browning of the flesh, it is hard to identify any physicochemical traits corresponding to quality criteria that apply to *D. rotundata* yams. According to Osagie (1992), the tuber starch content and the volume and gelatinization capacity of starch grains are crucial factors. Some cvs are probably distinguished by differences in genetic origin, but other factors also have a significant effect. African consumers claim that the cooking quality of yam cvs varies considerably according to the growing conditions, tuber size (for yams consumed in pounded form), harvesting time (relative to the end of the vegetative cycle) and end of dormancy (for stored yams).

Color problems associated with chip-based cooking do not seem to be related to other aspects of cooking quality. For instance, although some cvs (e.g. Kagourou, Tabande or Singou) from the Bariba region of Benin give an excellent whitish futu, they nevertheless produce a dark colored amala. This browning must therefore be due to processing. According to Chilaka *et al.* (2002) and Akissoé *et al.* (2003), peroxidase activity is at least partly responsible for the increase in phenol content during parboiling and drying of tubers for chip production.

It is hard to give a standard definition of cooking quality as it is perceived very differently by different consumers. This is largely due to differences in culinary practices and is changing in with urbanization. Quality perception is also complicated by regional preferences. While consumers in Côte d'Ivoire are strongly attached to the very distinctive flavor of cv Krengle, this yam has failed to find favor with the African diaspora in its industrially processed form. Different yam preferences are also noted in other African countries. The only *D. rotundata* yam currently meeting with unanimous approval from Côte d'Ivoire to western Nigeria is cv Kponan, because of its high organoleptic quality. It appears to be one of the major successes of domestication, particularly for making pounded yam, which is the finest food dish prepared from *D. rotundata* yams throughout most of West Africa.

Sexual features

D. rotundata yams, like all other Dioscoreaceae, are usually dioecious. Some cvs are male, others are female and a few include both male and female individuals. Rare cases of monoecism are known, but are often encountered with plants recently grown from seed (Hamon, 1987; Zoundjihèkpon, 1993). Monoecism might therefore be sign that the cultivated plant material is the result of recent domestication. *D. rotundata* yams have a highly variable flowering capacity due to several factors. Some are probably of genetic origin or linked to the length of time that the yams have been domesticated. Others are related to climate (rainfall quantity and distribution) and agronomic conditions (soil fertility, seed weight, planting density, time of emergence, staking practice, weed control). The latitude, and hence photoperiod, also has a substantial impact. Touré and Ahoussou (1982) observed that the flowering rate of some *D. rotundata* cvs in Côte d'Ivoire ranged from 70 to 86%, depending on whether the crop was grown in the southern or central part of the country. The photosensitivity of *D. rotundata* yams has been demonstrated by Okezie *et al*. (1993). It seems to be a general Dioscoreaceae trait that occurs in all wild West African species (Dumont, personal observation).

Some early-maturing cvs are male and others female, although Dansi *et al*. (1999) reported both male and female individuals in cvs of the Ahimon and Gnidou groups of Benin. The sex ratio can vary regionally. The two sexes are almost equally represented throughout Benin (Dansi *et al*., 1999). The predominance of male cvs was noted earlier in Côte d'Ivoire by Touré and Ahoussou (1982) and Hamon (1987), while the situation was almost reversed in Togo (Kassamada, 1992). Sixty percent of the 30 *D. rotundata* cvs studied in Cameroon by Dansi *et al*. (2001) were female. There appear to be at least as many females as males in early *D. rotundata* yams. A very different sex ratio will be noted later for their wild parents.

Few female cvs abundantly flower and fruit each year. Only four of the 33 female cvs observed over a 4-year period at the IITA station in Cotonou regularly produced a large amount of seeds (Dansi *et al*., 1999). The findings of Touré and Ahoussou (1982) suggest that flowering is probably more regular in savannah areas.

The inflorescences of male early cvs usually consist of a simple or compound raceme. Kponan and Laboko cvs form a specific group among early *D. rotundata* cvs as their individual inflorescences are not well developed and are sometimes even limited to a single flower. A similar marked reduction in flowering has never been observed in wild yams.

Mixed cvs are female, although cv Krengle (Côte d'Ivoire) includes 3% male individuals. It is also the only *D. rotundata* yam in which the fruit is sometimes entirely parthenocarpic (Dumont, personal observation). There are major differences in flowering and fruiting intensity among currently known mixed cvs. Abundant flowering occasionally occurs in cvs that are usually not very floriferous, suggesting inhibition rather than permanent impairment of sexual function.

The male sex is dominant in late-maturing *D. rotundata* yams throughout West Africa, in most cases with abundant flowering (Dansi *et al*., 1999). However, Zoundjihèkpon (1993) reported a marked reduction in flowering in the Frou and

Kroukroupa cvs of Côte d'Ivoire. At present, five female late cvs are known: Afoubessou (Côte d'Ivoire), Gnalabo and Kratchi (Benin-Togo) and two yet unidentified cvs (one in Burkina Faso, the other in Cameroon).

Studies conducted in Côte d'Ivoire (Zoundjihèkpon, 1993) suggested that fertility is low in late *D. rotundata* yams. With several male cvs, pollen grains appeared to be abnormally small. An *in vitro* experiment showed a mean pollen germination level of 5% for the (late) cv Gnan and no progeny was obtained from an attempted field cross between this cv and three (early) female partners chosen for their regular and abundant fruiting. Currently known late female yams produce morphologically abnormal (rugby-ball shaped) non-fruiting flowers. According to Zoundjihèkpon (1993), this is the result of a deficiency in the mechanism controlling corolla opening. Female sterility was reported by Abraham and Gopinathan-Nair (1991) in floriferous *D. alata* yams from India—only tetraploid individuals were fertile, while individuals with the highest degrees of ploidization (hexa- and octoploid) were always sterile. This phenomenon contrasts with the supposed sterility of late *D. rotundata* yams. Male cvs of the latter species always appeared to be tetraploid on the basis of the chromosome counts described in Section 2.3. Three of the five female cvs previously presented as being sterile (Afoubessou, Gnalabo, Kratchi) are also tetraploid, while the chromosome number of the two remaining cvs is unknown. The presumed sterility of late *D. rotundata* yams would thus appear to be unrelated to ploidization.

Several studies (Dumont and Vernier, 1997a; Baco, 2000; Okry, 2000; Mignouna and Dansi, 2002; Scarcelli, 2002; Vernier *et al.*, 2003) have reported that late cvs are rare or even non-existent among domestication products. This applies particularly to regions of West Africa where late cvs represent the bulk of yams cropped for commercial yam chip production or for storage as fresh tubers. This situation prevails in the northern half of Benin. Four surveys conducted there (Dumont, 1997; Dumont and Vernier, 2000; Vernier and Dossou, 2000; Baco, 2000) indicated that late cvs occupy 6? to 78% of the area cropped with *D. rotundata* yams. The marked contrast between the abundance of late yams in local agriculture and their extreme rarity among recent domestication products tends to confirm that late *D. rotundata* yams are highly sterile. This hypothesis was tested by the University of Abomey-Calavi (Cotonou) and CIRAD in 2002, in an experiment conducted in an isolated yam plot in Benin. A female cv Gnidou grown in the presence of four kokoro cvs (Yakanougou, Deba, Tabande and Kilibo) produced only three fruits, corresponding to less than 1% of the flowers that had formed. This result suggests that kokoro yams are actually sterile and that the rare fertilizations obtained were induced by pollen carried by insects from remote plants.

Other forms of *D. rotundata* yam generally also have a disturbed sexual function. In Côte d'Ivoire, a sterile female individual was noted among the progeny of a cross between early-maturing Sopere (male) and Lokpa (female) yams (Zoundjihèkpon, 1993). In the same country, the mixed cv Krengle occasionally produces spikes of sterile female flowers as well as spikes of fertile ones, and sometimes both types of flower grow on the same spike (Dumont, personal observation). Monoecious situations were noted by Zoundjihèkpon (1993) in young hybrids produced from early-maturing Ivorian cvs and she also emphasized that their fruit was often parthenocarpic. Daïnou *et al.* (2002) reported that a monoecious *D. rotundata* individual grown in an isolated situation in Benin produced fertile seeds. Monoecism was also observed by Dansi *et al.* (1999) in the

Ahimon and Gnidou cvs of Benin, which farmers regard as recently domesticated yams. In addition, the latter authors reported that setts of a single Ahimon tuber generated plants of different sexes. Zoundjihèkpon (1993) had previously reported sex reversal among juvenile hybrids of early-maturing Ivorian cvs. This suggests that sex is not always genetically determined and might be influenced by as yet unknown factors. It has been shown that heavy nitrogen fertilization of monoecious hemp (*Cannabis sativa*) masculinizes plants (Arnoux *et al.*, 1966).

Permanent sterility does not appear to be systematically linked to presumed interspecific hybridizations—there are other possible causes. Even though *D. rotundata* yams and their wild parents are autotetraploid, there is a high probability that unbalanced gametes will appear, resulting in sterility (Valdeyron, 1961). The presence of surplus chromosomes (Miège, 1952; Baquar, 1980; Zoundjihèkpon, 1993) might also have a negative effect on fertility. Occasional sterility is probably the result of physiological disturbances.

Cultivation requirements

Double-harvest (i.e. early-maturing) cvs require high organic, mineral and physical fertility and they are also severely damaged by nematodes. These cvs are therefore still mainly used in shifting agriculture, which involves long fallow periods. The environmental damage caused by this type of agriculture explains why there is some hesitation in adopting yams as a crop for sedentary intensified agriculture. These days, however, African farmers are often pushed to change their ancestral farming techniques. In western Burkina Faso, northern Côte d'Ivoire and northern Cameroon, the Senoufo, Lobi and Dourou peoples have all radically modified their cultivation techniques to enhance the profitability of growing double-harvest yams, by using cropping systems in which long-term natural fallow is virtually unnecessary. A major reduction in planting density is needed to grow cv Kponan (Dumont and Kouakou, in press), a variety that is highly appreciated for its cooking qualities but notorious for its high cultivation requirements. Shortening the fallow period often results in the use of poorer quality cvs, but this has little effect on the demand from urban consumers, who are primarily sensitive to the cost of the produce.

Single-harvest yams and, to a lesser extent, mixed yams are more tolerant of environmental conditions. A large proportion of these crops are therefore now included in cropping systems, thus substantially reducing the natural fallow period. In this new setting, the main constraint to yam cropping—rather than the much lower productivity—is the significant labor investment required for weed control (Vernier and Dossou, 2000).

Intrinsic productivity of the plant material

Intrinsic productivity is a criterion that clearly distinguishes double-harvest *D. rotundata* yams from single-harvest ones. Coursey (1976) reported that tubers of 60 kg can be found among double-harvest cvs for ritual ceremonies in Nigeria. An average yield of 26.5 t/ha was noted by Seignobos (1998) for cv Bakokae of northern Cameroon, which is widely grown on the Mbé Plain. An average yield of 23.9 t/ha was noted in three successive measurement campaigns conducted among farmers in northern Benin (Vernier and Dossou, 2000). Finally, 434 series of measurements carried out in Côte d'Ivoire during four crop seasons showed an average yield of 17.7 t/ha, with productivity peaks of nearly 50 t/ha (Dumont and Kouakou, in press).

Single-harvest *D. rotundata* yams have a much lower yield, which rarely exceeds 5 kg per plant. Productivity varies widely depending on the grower's objectives. Large tubers are required if the yams are to be marketed as fresh produce. The measured yield for this type of production hovers around 15 t/ha in northern Benin (Vernier and Dossou, 2000). If the yams are to be used to make yam chips, small tubers are an advantage and productivity can be improved by 25 to 50% by increasing the planting rate.

The known yields of mixed cvs are highly variable and strongly influenced by growing conditions. A yield of nearly 50 t/ha was recorded for cv Krengle in an irrigated cropping experiment in Côte d'Ivoire (IDESSA[1], 1986). In contrast, Vernier and Dossou (2000) reported a yield of less than 15 t/ha for cv Boni Oure grown on smallholdings in northern Benin, which is in line with the average yield recorded for this yam by the INRAB (Institut national des recherches agricoles du Bénin) station in the same region over the 1983-1995 period (INRAB, 1996).

All of the above yields are for traditional planting rates, i.e. ± 5,500 plants/ha for double-harvest cvs, mixed cvs and single-harvest cvs produced for storage as fresh tubers and ± 6,500 plants/ha for single-harvest cvs used to make dry chips.

Market adaptation

Each category of yam has been adapted to meet market expectations in a different way. The fresh tuber trade, which involves double-harvest cvs, has been boosted significantly by urbanization and to a lesser extent by the development of an export market in Ghana (Ghartey, 1994). *D. rotundata* yams (particularly cv Kponan) from Côte d'Ivoire, Ghana and Togo can also be found in some Paris markets and in Libreville (Gabon) (Dumont, personal observation).

There has been a boost in yam production from Guinea to Cameroon, resulting in regional specializations that sometimes involve people for whom yam-growing is a recent activity (Lobi of Côte d'Ivoire, Dioula of Burkina Faso, Dourou and some Fulani of Cameroon). This is an indication of the high profit potential sustained by strong market demand. Dumont *et al.* (1994) reported that growing double-harvest yams generated the highest earnings for a day's work in northern Cameroon. A similar situation was observed in Côte d'Ivoire (Bisson, 1989; Doumbia *et al.*, 2004), where 75% of the market supply involves double-harvest cvs (Touré *et al.*, 2003). This situation differs substantially from that projected by Onwueme and Charles (1994), who predicted that double-harvest *D. rotundata* yams would gradually be abandoned because they require high labor input.

Trade appears to have had less effect on the production of single-harvest cvs, at least in the short term. Single-harvest yams are harvested at the onset of the dry season, stored as fresh tubers and then marketed over the following months. Yam storage involves a risk

---

1. IDESSA: Institut des Savanes, now Centre Régional de Bouaké du Centre National de Recherche Agronomique (CNRA).

of loss for the producer and deterioration in the organoleptic quality of the produce, making it less competitive than other foodstuffs that are cheaper or easier to use.

The practice of boiling or frying yams has developed among urban populations in West Africa because it is less time-consuming. Production has responded to this demand by the adoption of cvs that purist consumers regard as being of questionable cooking quality (Kpassadjo and Sanata in Côte d'Ivoire, Gnidou in Benin) but less demanding and often higher yielding. It is likely that commercial agriculture in Africa will eventually abandon single-harvest *D. rotundata* yams in favor of *D. alata* cvs, which have the same advantages but are easier to store.

Small-scale processing of single-harvest yams into parboiled chips represents a significant change in commercial supply and sidesteps problems associated with the fresh tuber market. This technical change of direction dates back to the 1970s, when the development of Nigerian oil exports led to a major overall increase in food demand and prices (Igué, 1985). The main advantages of dried chips are significantly lower transport costs and a much longer storage life, with no increase in losses. Various research organizations have recently transferred yam chip technology to other parts of the subregion where this subsector had not yet developed (Bricas and Hounhouigan, 2000).

Wild yams *Dioscorea praehensilis* Benth and *Dioscorea abyssinica* Hoest ex Kunth

General

These two yams belong to the Enantiophyllum Uline section, which N'Kounkou (1994) considers to be one of the least evolved of the eight subfamilies (sections) of Dioscoreaceae found in Africa, and therefore likely a major source of variability. Two dynamics separately or jointly appear to be responsible for this variability. From 1939, Burkill (and Chevalier, apparently as early as 1936) suspected that some members of the Enantiophyllum section were interspecific hybrids. A number of authors also noted that ploidization occurred in several species of this section, notably in cultivated yams (Miège, 1952; Martin and Ortiz, 1963; Baquar, 1980; Essad, 1984; Zoundjihèkpon *et al.*, 1990; Abraham and Gopinathan, 1991; Hamon *et al.*, 1992; Gamiette *et al.*, 1999; Dansi *et al.*, 2000b; Dansi *et al.*, 2001; Camara, 2001).

These two dynamics appear to be determining factors in an ongoing evolutionary process, as highlighted by the present incomplete interspecific separation with respect to biological functioning and morphological traits. Hamon (1987) showed that the *D. mangenotiana* species included individuals with very different vegetative organ renewal frequencies and N'Kounkou (1993) considered that they belonged to two distinct species. The one with annual vegetative organs would be *D. mangenotiana stricto sensu* and the other *D. baya* De Wild. It is also hard to accurately define the morphological boundary between *D. praehensilis* and *D. abyssinica*. Moreover, the latter is often regarded as the same species as D. *lecardii* De Wild, which in turn is poorly separated from *D. sagittifolia* Pax.

There seems to be high instability in the Enantiophyllum section, unlike other Dioscoreaceae. This probably provides many opportunities for domestication wherever these yams are found. The Enantiophyllum section has produced the *D. alata* species in Asia, and *D. cayenensis* and *D. rotundata* in Africa. These domestication products, with their high cv diversity, account for virtually all yam production worldwide.

Most criteria for the biological separation of *D. praehensilis* and *D. abyssinica* are based on taxonomic studies of herbarium plant material. Until recently, little effort has been made to collect information from African farmers, even though they are often very well informed about the local plant environment. There have been no population-scale studies of the wild species concerned either, so the range of variability of the two species and any connections between them are still largely unknown. These factors explain the difficulty in separating *D. abyssinica* from *D. praehensilis*. The two species nevertheless have several traits in common.

Both usually produce only one tuber per plant, but it is not unusual to come across double-tuber *D. abyssinica* plants. The tuber and aerial vegetative organs are renewed annually, with the tuber taking over to ensure the plant's survival in the dry season and subsequently breaking down completely. *D. praehensilis, D. abyssinica* and the domesticated forms are actually both perennial as well as geophytic and tropophytic species (Miège, 1952), since their biological flexibility enables them to adapt to a climate with alternating wet and dry seasons.

The sex ratio is heavily weighted in favor of male plants. In a study of yam specimens conserved in 10 national or university herbaria, N'Kounkou (1993) noted a male/female ratio, among flower-bearing plants, of 33/1 for *D. praehensilis* and 7/1 for *D. lecardii* and *D. sagittifolia*, both of which are taxonomically close to *D. abyssinica*. The same inventory also showed a heavy preponderance of male plants in the other wild species of the Enantiophyllum section. This male preponderance might be explained by the biological characteristics of Dioscoreaceae. In the wild, the plants are often very scattered and pollinated by insects (entomophily), especially by *Larothrips dentipes* thrips (Pitkin, 1973; Zoundjihèkpon, 1993), whose population size is determined by climatic conditions. Surplus pollen is essential to ensure a sufficient supply in this uncertain setting.

The under-representation of the female sex means that allelic diversity is limited. Female plants produce several dozen to several thousand flowers, each with six ovules, so each plant can be pollenized by a wide diversity of male parents. The products of these fertilizations are half-sibs as they all have the same maternal genetic heritage. Applying this reasoning more generally, Atlan and Gouyon (1994) showed that the less represented sex (female) has a reproductive advantage as it has a higher probability of passing on its genes.

D. abyssinica and *D. praehensilis* also have other general traits in common. Firstly, their preferred ecosystem is a regenerating plant environment. Both species are very dependent on fallows or windfall areas, where they grow while the climax plant populations become re-established.

Secondly, the aerial architecture of *D. abyssinica* and *D. praehensilis* is typical of wild yams overall. The voluble stems of these extremely heliophile species grow to a considerable height before branching as they need to rise above the supporting shrub vegetation before putting out leaves and flowers.

Lastly, both species are extremely polymorphic. This variability is generally linked to the age of the plant, which can be considered on two levels. On one level, it is the duration of the life of the vegetative organs and tuber, which is short as both parts are renewed annually. On another level, it is the genotype age, which corresponds to the number of annual vegetative cycles since the plant grew from seed. The genotype and phenotype ages are the same since the plant is a perennial. Domestication upsets this synchronism, as will be seen later.

Some traits vary during the annual vegetative cycle. The extrafloral nectar and cataphylls disappear gradually, and later the stem loses its waxiness. Leaf shape and size change as the plant grows. Lastly, the stem color becomes uniform as senescence approaches.

Other morphological traits are modified as a result of interannual variations. The very elongated leaf shape and the waffled appearance and wine color of the leaf blades are linked to the juvenility of the genotype. The red stem color and lanceolate *D. lecardii*-type leaves of *D. abyssinica* seem to only occur in recent genotypes. The morphological diversity of a genotype is generally reduced with aging. Could this represent a stabilizing effect of natural selection? If so, there is probably a corresponding loss of genetic diversity.

Contrary to these numerous common points, there are several fundamental criteria that separate the two species *D. praehensilis* and *D. abyssinica*.

Dioscorea praehensilis

D. praehensilis has a very wide geographical distribution in Africa. It is found throughout the central part of the continent and is also widely distributed in tropical Africa. In the northern hemisphere it can grow beyond the 8th parallel, whereas in the southern hemisphere it has been encountered as far south as Mozambique (N'Kounkou, 1993). So far this species has been strictly regarded as a forest yam. It grows abundantly in post-fire regeneration areas amongst dense communities of semi-deciduous trees, which Mangenot (1955) defined as mesophyll forests. It is common in the bimodal rainfall zone, but in drier climatic conditions it takes refuge in the rare vestiges of mesophyll forest that have survived the combined effect of annual fires and anthropogenic pressure.

The species *D. liebrechtsiana* De Wild, which ranges from Central Africa to forest Cameroon, is morphologically very close to *D. praehensilis*. Its main distinctive criteria are the strongly auriculate base of the petiole and the slight spininess of the stem. The Zande peoples of the Democratic Republic of the Congo (formerly Zaire) refer to both yams by the same name (N'Kounkou, 1993), and Wilkin (2001) does not make a distinction between them. *D. liebrechtsiana* De Wild could thus be a variant of *D. praehensilis*. This would be a first illustration of *D. praehensilis* diversity.

Yams currently classified as members of the *D. praehensilis* species are clearly divided into two distinct groups. Miège (1952) separated the *D. praehensilis* yams of southern Côte d'Ivoire into two types according to the morphological traits of the tuber. Both morphotypes have also been found in the Central African Republic (Hladik *et al.*, 1984). They are mentioned (as *D. rotundata* yams) in a publication by Chikwendu and Okezie (1989) and again by Hamon *et al.* (1995) in their book on the wild yams of West Africa. The two morphotypes will hereafter be referred to as Dp1 and Dp2.

Bariba farmers of northern Benin identify Dp1 ('Sonkotouné' in the vernacular) on the basis of a number of traits in the aging genotype. The stem is dark-colored, it has a basal diameter of over 1 cm and its lower part has 2-4 cm long cataphylls with fleshy sheathing lobes. The leaf is slightly cordate and has no marginal undulations. The tuber is large (sometimes weighing over 5 kg) and irregularly lobed, its thickness increases with length, often reaching 25 cm at its distal part, and the flesh is speckled with mauvish spots (except in the neoformed part). The tuber head is surmounted by extremely thorny roots that form a permanent protective crown at the soil surface. Finally, the male flowers have a very characteristic scent. The only West African species to share this trait is *D. hirtiflora* Benth. This similarity confirms the phyletic proximity of the Enantiophyllum and Asterotricha sections, as suggested by Burkill (1960), Hamon (1987), N'Kounkou (1994) and Schols *et al.* (2001). The cladogram constructed by Terauchi *et al.* (1992) on the basis of the chloroplast DNA characteristics shows that the oldest genome of the Enantiophyllum section in Africa corresponds to the *D. praehensilis*, *D. abyssinica*, *D. rotundata* and *D. cayenensis* species. The fact that the first three have the same genome does not mean they have the same age but rather that they belong to the same evolutionary line and may actually have appeared successively. The ancestral form would be Dp1, because of its phyletic proximity to the Asterotricha section and its extremely wide geographical distribution in areas where *D. rotundata* and *D. abyssinica* are not found.

The neoformed part of the Dp1 tuber is edible but not very appreciated. It was formerly eaten in Benin during food shortages. It is criticized for its excessively fibrous texture and bitter taste. However, it still seems to be gathered in southwestern Ethiopia, as Hildebrand (2003) reported that wild yams belonging to the Enantiophyllum section and protected by a cluster of spiny roots were harvested locally.

The Dp1 yam grows in mesophyll forests but is endangered as a result of the disappearance of its biotope. Some patches of mesophyll forest still exist in humid parts of Africa, but in arid areas with just one rainy season only remnants are found, in the form of sacred forests. The Dp1 yam is rare or extinct in some areas where savannah flora now thrives—this is probably a historically recent phenomenon. Genes from this morphotype may still circulate locally in existing wild yam populations or among *D. rotundata* yams derived from long domesticated cvs.

The Dp2 form of *D. praehensilis* differs substantially from Dp1 except for the characteristic scent of the male flowers. Moreover, ecotype variations sometimes link it morphologically to the *D. abyssinica* species, which is examined in the following section. The stem is light-colored, usually no more than 0.6 cm in diameter and has short (< 4 cm) cataphylls that do not develop large lobes. At the leaf level, two reliable distinctive signs (in aging genotypes) are the wide opening of the foliar lobes and the regular undulation of the edge of the leaf blades. The tuber is cone-shaped, seldom more than 30 cm long, usually weighing under 3 kg and bearing sturdy thorns, although this appears to be an irregular trait (Tostain, 1998; Tostain and Daïnou, 1998). The flesh is white or pale yellow and often partly purple at the head of the tuber.

Dp2 yams are edible and highly esteemed. Large quantities are gathered in forest areas. In West Africa, it is harvested mainly at the onset of the rainy season (late January to March) since it is easier to locate the tubers once they have germinated. Germination also improves their cooking quality, as the bitterness is masked by partial conversion of the starch into sugar (Osagie, 1992). If the yams are harvested at the end of the dry season, the tuber heads can be replanted, thus ensuring the sustainability of this food resource. However, the yams are commonly gathered much earlier, and large quantities are sold from July to September in the markets of the eastern forested area of Guinea (Dumont, 1993).

Lastly, in West Africa below latitude 8° N, the Dp2 form of *D. praehensilis* is the preferred raw material for domestication, leading to the creation of *D. rotundata* cvs. This practice is very widespread in southern Benin, where Gnidou cvs are domesticated. This trend probably also applies throughout forested regions of Africa, but this needs to be qualified. Some farmers in Benin and Nigeria say they start domesticating *D. rotundata* cvs from pieces of Dp1 tuber (Vernier, personal observation). In addition, this information has been reported by Chikwendu and Okezie (1989) and confirmed by Scarcelli (2002). In Central Africa, the wild yam *D. liebrechtsiana* might also be domesticated to produce *D. rotundata* cvs. This is supported by the fact that *D. liebrechtsiana* has a highly auriculate petiole—a trait noted in one of the nine varietal groups in which Martin and Rhodes classified 97 *D. cayenensis-D. rotundata* cvs collected in Africa and the West Indies (in Degras, 1986).

The Dp2 morphotype is rare in the monomodal savannah above latitude 9° N and grows only in the most humid microclimates. We have never found it in stable remnants of mesophyll forest, but it spreads very rapidly (from seed) when this ecosystem degrades—where it can thrive until the final stage of wooded savannah. As the degradation process advances, the Dp2 morphotype gradually converges morphologically with that of *D. abyssinica*. There is still evidence of this morphological transformation in yams growing in the Upper Oueme catchment of Benin.

The Dp2 form readily diversifies into different ecotypes while maintaining a certain amount of genetic diversity. More technically, it has a wide scope of adaptability (Binder, 1972), the full range of which seems to prevail in the southernmost part of the monomodal savannah. It is therefore not surprising that Dansi *et al.* (1999a) regarded *D. praehensilis* as the starting-point for several *D. rotundata* cvs derived from savannah flora in which often only *D. abyssinica* yams are visible. In the same ecological area, the origins of *D. praehensilis* yams that colonize the degrading remnants of mesophyll forest and of those that colonize regenerating mesophyll forest communities are also clear.

The Dp2 morphotype is known as the Bayere yam in the West African area between eastern Côte d'Ivoire and Benin which is under a bimodal rainfall regime. Farmers in Togo make a distinction between the forest Bayere yam and the form that grows in the savannah biotope after degradation of the forest ecosystem (Kassamada, 1992). The first is a typical Dp2 yam and uses large forest trees for support. The second type could occur in two forms. One would be a Dp2 yam whose morphotype has been modified as an adaptation to the savannah ecosystem. The other would be derived from natural seedlings of *D. rotundata* yams domesticated from *D. abyssinica* and *D. praehensilis* yams in more northerly regions and then cultivated under the bimodal rainfall climate. In support of the latter hypothesis, Dansi *et al.* (1999a) identified three groups of locally domesticated cvs (Ourtchoua, Porchebim, Mondji) in the southern half of Benin as being morphologically linked to *D. abyssinica*, despite the fact that the latter species is not endemic to the region.

As the yam regarded as *D. praehensilis* includes two quite different morphotypes, the relations between them need to be examined. H.M. Burkill (1985) first proposed that there may be two genetically separate species—Dp2 would be *D. praehensilis* while Dp1 would be related to the old species *D. odoratissima* Pax. However, several botanists (Baker, 1898; Durand and Durand, 1909; Raadts, 1984; N'Kounkou, 1993) classified the latter as a *D. praehensilis* yam. Based on the anatomical characteristics of the leaf midrib, Edeoga and Okoli (2001) advocated that *D. liebrechtsiana*, *D. praehensilis* and *D. odoratissima* be included in the same taxonomic group. This proposal was further confirmed by Wilkin (2001) on the basis of other scientific arguments. For the moment, Dp1 and Dp2 will be regarded as morphological variants of the *D. praehensilis* species, as defined by Miège (1968). Anthropogenic pressure could possibly account for this differentiation. Dp2 might be a transformation of Dp1 induced by repeated gathering and/or the adaptation of the plant material to simple cultivation techniques (which we will discuss later). This was also suggested by Dounias (1996), who encountered Dp2 in forest areas of Cameroon and concluded that it had been introduced there by man. A. Hladik and Dounias (1996) then investigated the correlation between the abundance of *D. praehensilis* yams and the presence of hunter-gatherers. The latter might well be the decisive factor. The abundance of wild yams in an unfavorable ecosystem might be the result of long-term food resource management by human groups. The above authors believed that this management process involved the farming of wild yams in their natural environment. We should also mention that open areas left behind after human encampments and associated nearby cultivated areas represent a suitable biotope for the proliferation of *D. praehensilis*. In addition, Dounias (1996) reported that this yam grows along forest trails, so its distribution clearly overlaps areas where human activities are focused.

Three examples highlight the genetic compatibility of *D. praehensilis* and *D. rotundata*. A case of successful hybridization of these two yams was reported by Akoroda (1985). The fertilization of several *D. rotundata* cvs by *D. praehensilis* was demonstrated by Zoundjihèkpon (1993) using enzymatic markers. A female cv Krengle plant sown on a path in the mesophyll forest of Côte d'Ivoire also apparently hybridized with the Dp1 form of *D. praehensilis* (Dumont, personal observation). The latter assumption is based on the very spiny roots of the progeny.

No exhaustive studies have been conducted to date on the genetic diversity of *D. praehensilis*. Most information on this subject is derived from studies carried out by Hamon (1987) in Côte d'Ivoire, Scarcelli (2002) in Benin and Tostain *et al.* (2002) in Guinea, Togo, Benin and Cameroon. Enzymatic markers were used in the former study and molecular markers in the other two.

The *D. praehensilis* yams of Côte d'Ivoire were separated into two genotype classes. The same result was obtained with five plants from northern Benin, but a study of 18 individuals from southern Benin revealed three classes. Neither study highlighted a parallel between the genotype groups revealed and the two *D. praehensilis* morphotypes defined above (Dp1 and Dp2).

The results of the four-country study of Tostain *et al.* (2002) separated the 46 *D. praehensilis* yams examined into four groups on the basis of a correspondence analysis. The genotypes were generally grouped by geographical origin, but this needs to be qualified. While genetic bridges were noted between the plant material from Guinea, Togo and Benin, *D. praehensilis* yams from Cameroon were genetically isolated from those of the other three countries. This suggests that the species consists of two groups, one specific to West Africa and the other to Central Africa. The former is in the 'civilization of the yam' belt, while the latter has been subject to much weaker domestication pressure. This difference may well establish the dividing line between the two *D. praehensilis* populations.

Dioscorea abyssinica

D. abyssinica was long regarded as being the same species as *D. togoensis* (Miège, 1952[2]; Waitt, 1965). H.M. Burkill (1985) did not separate *D. abyssinica* from *D. sagittifolia*, while implicitly classifying the latter as a *D. lecardii* yam. *D. abyssinica* grows mainly above the Equator in sub-Sahelian Africa. It is abundantly distributed from the western edge of this region to Cameroon (Jacques-Félix, 1947; Miège, 1968; H.M. Burkill, 1985; Dumont *et al.*, 1994) but has not been encountered in forested areas of Central Africa (Hladik *et al.*, 1984; N'Kounkou, 1993). More generally, it has never been reported south of the Equator. On the other hand, H.M. Burkill (1985) and Gebre Mariam and Schmidt (1996) reported its presence in East Africa, especially in Ethiopia. As early as 1949, Chevalier thought that the 'true' *D. abyssinica* originated from Abyssinia (now Ethiopia) and Kenya. In contrast, certain other wild yams existing outside of East Africa might well be 'false' *D. abyssinica* yams. We will discuss this idea later.

D. abyssinica is strictly a savannah yam, and it appears to prefer a monomodal rainfall regime. It therefore grows mainly in the climatic belt roughly ranging from latitudes 8° to 12° N.

In *Flora of West Tropical Africa*, Miège (1968) indicated that *D. abyssinica* is distributed throughout the area in which *D. rotundata* yams were domesticated, while placing *D. lecardii* along the northern edge of this area. The limited geographical distribution of *D. lecardii* suggests that it was able to survive as a separate species only on the fringes of a particular initial area of occupation, which was once probably far more extensive.

D. lecardii and *D. sagittifolia* yams meeting the morphological criteria established by Miège (1968) are now rarely found among the savannah flora of West Africa. Their identification is often doubtful, particularly when it concerns herbarium plant material. *D. lecardii* is usually classified as a *D. abyssinica* yam. Morphotypes clearly corresponding to the *D. sagittifolia* species have been found in two countries. One was noted in a relict savannah within the forest area of southwestern Côte d'Ivoire (Dumont, personal observation) and seven others were found in sacred forests in southern Benin (Allomasso, 2001). According to Miège (1968), *D. sagittifolia* yams have also been collected in Senegal, Sierra Leone, northern Côte d'Ivoire and southern Burkina Faso, but these identifications were based on herbarium specimens.

Further eastward, *D. lecardii* and *D. sagittifolia* have distinctive individual characteristics and they become abundant from Cameroon, i.e. virtually from the easternmost edge of the *D. rotundata* distribution range, and onward (Dumont *et al.*, 1994). Jacques-Félix (1947) found that *D. lecardii* was cultivated in northern Cameroon and classified it as cv Dem of the Dourou region. We consider that the Tii and Ngan cvs of the same region were also domesticated from *D. lecardii* on the basis of their morphological traits and particularly the small tuber size. The same is probably true of the cultivated 'Haab' yams of Chad (Dumont, personal observation).

A double hypothesis can be put forward on the basis of these ideas. In Cameroon and further eastward, *D. abyssinica, D. lecardii* and *D. sagittifolia* would be a single species that has been separated into relatively distinct ecotypes through large-scale adaptive

---

2. Miège recognized *D. togoensis* as a separate species in the 1968 edition of *Flora of West Tropical Africa*.

polymorphism. This has been partially verified experimentally (Dumont, personal observation). A yam with *D. lecardii* characteristics (according to Miège, 1968) collected in northern Côte d'Ivoire in 1991 expressed the foliar traits of *D. sagittifolia* after it was cropped in the southern part of the country. The situation seems to differ in West Africa, where *D. rotundata* yam cultivation has long accompanied the transition of forest to savannah. In this region, *D. abyssinica* yams would generally correspond to a cultigen population resulting from domestication pressure on endemic *D. lecardii*, *D. sagittifolia* and *D. praehensilis* yams. This takes us back to the idea of Chevalier (1949), who made a distinction between East African *D. abyssinica* yams and the others. Moreover, variability has been specifically noted in *D. abyssinica* yams in the part of Cameroon adjacent to the *D. rotundata* growing area. Some foliar traits found in these yams, the orange coloring of the cataphyll acumen and the large leaf blade area have never been reported in West Africa (Dumont *et al.*, 1994). If West African *D. abyssinica* yams are cultigen populations, they cannot have produced many *D. rotundata* cvs in areas where strong domestication pressure is historically recent, as clearly illustrated in Côte d'Ivoire.

A number of morphological traits are specific to the West African forms of *D. abyssinica*. The light stem coloring, the concentration of spines at the axil of cauline branches and the absence of undulatory deformations on the edge of leaf blades are the most reliable identification criteria in aging genotypes. The tuber is usually cylindro-conical in shape, with the widest diameter at the top. It ranges from 20 to 80 cm in length and weighs between 0.5 and 1.5 kg. The flesh can be pure white to cream-colored, sometimes with subcutaneous purple pigmentation at the head. The bitterness of the flesh varies, apparently due to the influence of numerous factors (individual plants, age, environmental conditions, harvesting time). *D. abyssinica* yams are edible and are still highly sought after by savannah peoples since they are mature several weeks before the earliest *D. rotundata* yams (Baco, 2000).

In both regions of northern Benin studied by Dumont and Vernier, Bariba people refer to *D. abyssinica* yams as 'dika' (Dumont and Vernier, 1997a). This name encompasses four morphotypes. For simplicity, these are classified in two groups, i.e. Da1 and Da2.

Da1 includes the so-called 'dika sinrou konsi' and 'dika kpika', meaning 'yam with the end that goes rotten' and 'white yam' (which has light-colored leaves). Both have long tubers that are not much more than 5 cm in diameter at the head.

Da2 includes 'dika wonka' and 'dika guea' (literally 'black dika' and 'true dika'). More generally, these names refer to a yam with the dark leaves of a double-harvest *D. rotundata* yam and a relatively large tuber terminating with large fingers. The latter are particularly cherished by yam gatherers.

The Da2 form is well established in savannah flora (including natural fallows) close to *D. rotundata* crops. These Da2 yams can be plants derived from seeds that have escaped from crop fields (zygotic migration) or products of cross-pollination between wild dika yams and cultivated *D. rotundata* yams that are still sexually functional (gametic migration). Da2 definitely serves as starting material for *D. rotundata* yams in savannah areas. Once removed from its natural environment and grown agriculturally, the Da2 yam is called 'Tamdika' (from tam, plural tassou, the Bariba word for yam) when domestication is under way, and 'Tamdwe' when it has been adopted by farmers for cultivation with their double-harvest *D. rotundata* yams. Only vegetative reproduction is used in this transformation process.

The Da1 form occurs in sympatry with Da2, but its range extends over a much wider area to the north. In Benin, it is abundant up to the Niger River Valley, above latitude 12° N, and also grows up to the same latitude in Burkina Faso (Dumont, 1982). Several morphological traits are modified in Da1 yams growing in this northern region (Dumont, 1982). The stem spininess and fruit size are reduced, while the length-to-width ratio of the leaf is increased. These variations could be interpreted as atavistic throwbacks to the *D. lecardii* species, since *D. abyssinica* yams growing in West African savannah areas with a monomodal rainfall regime are suspected to have originated from *D. lecardii*.

The Da1 morphotype thus grows much beyond the area in which *D. rotundata* yams are cultivated. In Benin, yams are not planted much above latitude 10° N. This also seems to be the pattern throughout most of West Africa, except in the Republic of Guinea. In all cases, the 1,000 mm annual isohyet seems to be the threshold below which yam-growing becomes risky in rainfed cropping systems on normally drained soils. Da1 yams growing beyond this limit might be remnants of *D. rotundata* cultivation that were once cultivated further north. We will examine this hypothesis later.

Several scientific studies have shed light on the genetic diversity of *D. abyssinica*.

Hamon (1987) separated the *D. abyssinica* yams of Côte d'Ivoire into two classes on the basis of an isozyme analysis, although the study was limited to 18 plants of the same maternal origin.

Ramser *et al.* (1997) placed *D. abyssinica* in a genetically intermediate position between *D. praehensilis-D. liebrechtsiana* and *D. rotundata* (using split decomposition) on the basis of four types of molecular markers.

AFLP analysis revealed a contrasted situation in Benin (Scarcelli, 2002). Fifty-eight *D. abyssinica* plants from northern Benin were distributed in eight genotype classes but there was no agreement with the classification used by Bariba farmers. Nine plants from southern Benin were separated into three classes. It was shown that cv Gnidou is close to *D. abyssinica* even though its wild ancestor is *D. praehensilis* (Dansi *et al.*, 1999). Wild yams identified as *D. abyssinica* would thus actually be *D. praehensilis* yams that adapted to the forest to savannah transition or that had escaped from cultivated fields in the form of seeds. In addition, the dendrograms constructed by Scarcelli (2002) generally show genetic continuity between *D. praehensilis* and *D. abyssinica*. Two genotypes of *D. praehensilis* (Dp1 morphotype) are distributed among the *D. abyssinica* yams of northern Benin while two genotype classes in southern Benin include both *D. praehensilis* and *D. abyssinica* yams.

Using AFLP analysis, Tostain *et al.* (2002) constructed a dendrogram that separates 44 *D. abyssinica* plants from Benin, Togo and Guinea into three groups. One contains virtually all the Guinean genotypes. A second consists mainly of Beninese genotypes. The third group contains genotypes collected in Benin and Togo. The genetic diversity of *D. abyssinica* would appear *a priori* to be geographically structured, particularly in Guinea. It should be remembered, however, that *D. abyssinica* has been domesticated in Guinea only in recent times whereas, as discussed later, domestication appears to be a very old practice in Benin and Togo. This difference is perhaps responsible for the genetic discrepancy revealed in the Guinean plant material.

Analysis of chloroplast DNA of 147 Beninese *D. rotundata-D. abyssinica* yams by Chaïr *et al.* (2005) revealed five *D. abyssinica* individuals separated from the others by a different haplotype. *D. abyssinica* yams in Benin would thus appear to consist of two distinct species. In this setting, it is important to note that Hamon (1987) reported that *D. mangenotiana* is occasionally used to domesticate *D. rotundata* yams in Côte d'Ivoire. Future research will perhaps lead to the discovery of other haplotypes among *D. abyssinica* yams. Likely candidates are the wild parents of *D. rotundata* yams cultivated in pits in Cameroon and those of the cv Singou from the Bariba region of Benin — the yam with *Hymenocardia acida* (Euphorbiaceae)-like leaves.

Addendum: *Dioscorea togoensis* Knuth

This species has so far not been discussed. In savannah regions, this wild yam of the Enantiophyllum section is closely sympatric with *D. abyssinica*, a species considered to be one of the parents of *D. rotundata*. Curiously, there is also considerable overlap between the distribution areas of these three yams in West Africa and all terminate in Cameroon (figure 4). Lastly, *D. togoensis* has been confused with the *D. praehensilis*, *D. lecardii* and *D. sagittifolia* species on a number of occasions in taxonomic studies of herbarium specimens (N'Kounkou, 1993). This suggests a possible phyletic relation between *D. togoensis* and wild yams that are more or less closely related to *D. rotundata*. However, this view is certainly not supported by current scientific results. When measured by flow cytometry, the *D. togoensis* genome appears to be significantly smaller than that of *D. abyssinica* (Hamon *et al.*, 1992). RFLP analysis of chloroplast DNA revealed that the *D. togoensis* genome is not only unique but also the youngest among the African Enantiophyllum (Terauchi *et al.*, 1992). *D. togoensis* therefore does not seem

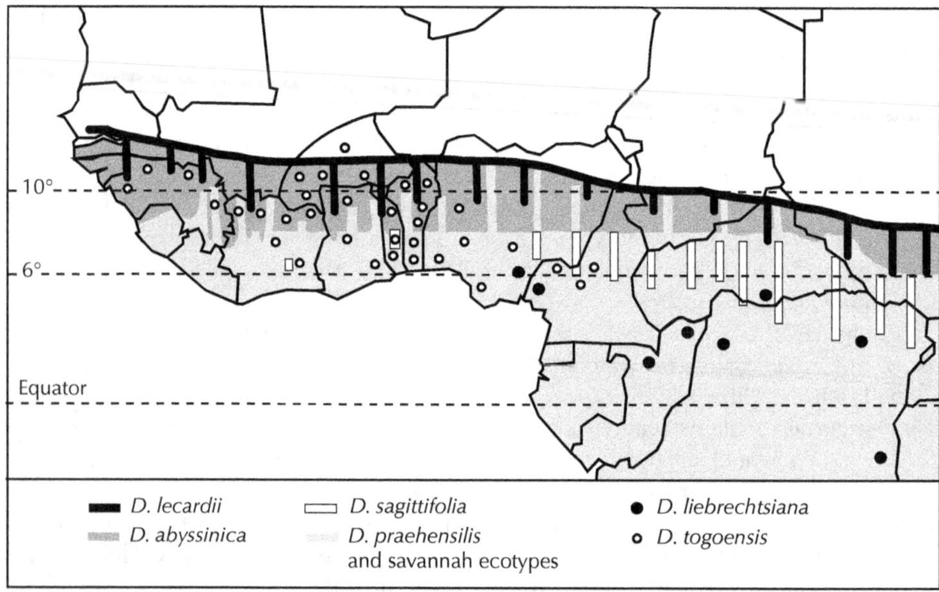

Figure 4. Geographical distribution of wild yams of the Enantiophyllum section with annual tubers. Extrapolated from information provided by Miège (1963), H.M. Burkill (1985), N'Kounkou (1993) and personal observations of the authors.

to be phyletically related to *D. abyssinica* and *D. praehensilis*, and any contribution to the phylogenesis of *D. rotundata* is inconceivable in the present state of our knowledge. For the moment, *D. togoensis* will be considered as the result of a fine speciation process to adapt to the savannah ecosystem. Future studies should lead to a better understanding of the relations between the African wild yams of the Enantiophyllum section. In this setting, we felt that it was essential to mention the *D. togoensis* species.

Phenomena that could explain the variability of wild and domesticated yams

Climatic disturbance

The Dioscoreaceae probably appeared 130 to 65 million years ago during the Cretaceous period, at the same time as other angiosperms. According to Barale and Lemoigne (2000), all plant families known today existed at that time. The genetic isolation of the African Dioscoreaceae dates back to the Miocene (60 to 20 million years ago) when the desertification of what is now southwestern Asia occurred (Coursey, 1976). The diversification of the Enantiophyllum section in Africa certainly took place less than 40 million years ago—no yams of this family apart from the introduced *D. minutiflora* exist in Madagascar (Burkill and Perrier de la Bathie, 1950), which separated from the African continent during that period (Jaeger, 2001). Hence, in terms of the evolution of the plant kingdom, African yams of the Enantiophyllum section, including the putative parents of *D. rotundata*, would *a priori* be relatively recent. This perhaps explains the impression of incomplete speciation they sometimes give and the large reserve of variability some of them still appear to have.

Because of their long history, African yams have survived through many climate changes and associated glacial-interglacial cycles that our planet has experienced over the last 2.5 million years. The Late or Upper Dryas was the most recent cold period of global scale affecting northern Equatorial Africa—it occurred 10,500 years ago (Lézine, 2000) and during this period severe drought led to a slight retreat of the forest in Central Africa. The idea of temporary forest decline in prehistoric Africa had already been put forward by Schnell (1971).

According to Lézine (2000), the end of the Late Dryas resulted in climatic warming and higher rainfall. Under these favorable conditions, subtropical vegetation would have spread northwards to the Sahara and desert areas would have virtually disappeared. "The conditions were right for a tremendous development in human activities, animal husbandry and agriculture, as abundant rock carvings and paintings testify." Coursey

(1976) believed that this situation was reversed 4,000 to 5,000 years ago, leading to the gradual climatic drying of northern Equatorial Africa. This phenomenon continues today and over the last few centuries its effects have been accentuated in humid West Africa as a result of increased population pressure.

Yams may well have followed the northward advance of tropical vegetation at the end of the Late Dryas, at least to a certain extent, and then retreated as the climate became more arid. We have no scientific arguments concerning the scale of the presumed migrations. Paleobotanical research, particularly in the field of palynology, may have generated some information on this subject, but no search of this literature has been conducted yet. There may well be fossil evidence of African yams from prehistoric times. In Panama, starch grains from Dioscoreaceae have been found adhering to 8,000 year old Stone Age tools (Piperno and Holst, 1998).

Relictual woodland flora found in West Africa up to latitude 11° N indicates that the area was once forested, but the savannah has spread 400 to 600 km further south. Parts of this ecologically transformed area are cohabited by *D. rotundata* yams and their wild parents.

Conversion to savannah exerts a strong evolutionary pressure. Our previous observations and the work of Scarcelli (2002) highlighted the development of a *D. praehensilis* ecotype adapted to the savannah following the recent degradation of mesophyll forest in Benin. A more fundamental effect is suspected, which would have occurred over a long period as a result of gradual climatic drying, i.e. the *D. praehensilis* species, the only forest-dwelling parent of *D. rotundata*, may have lost some of its adaptability. The usual effect of such a phenomenon is a loss of genetic diversity and a shift in the homeostatic equilibrium point (Binder, 1972). A population develops with its own particular traits, thus morphologically and genetically distancing it from the original plant material. This may well be how *D. lecardii* arose, but adaptive drift does not appear to have led to reproductive isolation (speciation) as the *D. lecardii* population introgressed into *D. abyssinica* in the *D. rotundata* domestication area. The fact that Chevalier (1920 and 1936) regarded *D. praehensilis* and *D. lecardii* as ancestors of cv Soussou of northern Benin does not contradict the hypothesis that they are two forms of the same species.

As far as cultivated yams are concerned, conversion to savannah may well have affected the plant material used and the geographical range of the cropping area. From Guinea to northern Cameroon, *D. cayenensis* yams and the taxonomically related hexaploid forms appear to be the surviving traces of a yam that was once widely grown but then marginalized by the development of *D. rotundata*. This change in plant material appears to have been the process which gave rise to the 'civilization of the yam' (Miège, 1952)—this pattern could be explained in two ways. Firstly, because of their very short dormancy period, *D. cayenensis* yams could not ensure food security at times when rainfall declines and population growth increases, thus prompting the need for food reserves (Burkill, 1918; Hutchinson and Dalziel, 1931; H.M. Burkill, 1985), Secondly, there were no double-harvest *D. cayenensis* cvs (Hamon, 1987), so production could not be temporally extended.

These days, as we have already seen, yams are not grown much above latitude 10° N. However, some small yam cropping areas are found beyond this limit, usually on hydromorphic soils where early emergence is possible and there is a regular supply of water at the onset of the dry season. Their northernmost location is on the Pilimpikou Plain in central Burkina Faso (Dumont and Hamon, 1985), at around latitude 13° N, where a few hectares of the male cv Bolgo Nyu (n.b. Bolgo = hydromorphic soil, Nyu = yam), hexaploid, according to Zoundjihèkpon (1993) crops are grown. This Pilimpikou yam is the only known case of monocropping based on this type of yam in West Africa. About 50 ha were estimated to be under cultivation in the 1970s (Dumont, 1980) but the area declined by 90% over the next 2 decades and was finally restricted to a single village, i.e. Arbolle (Goudou-Urbino *et al.*, 1996). An analysis of the reasons for this severe decline are beyond the scope of our study. It should nevertheless be pointed out that the last traces of an age-old form of agriculture are in the process of disappearing. According to Kassamada (1992), the same cv survives under a different name, i.e. Kpeyou, among the Kabye and Kotokoli peoples of Togo. Hence, a plant material that once had an extremely wide distribution area is now reduced to relict areas.

At a lower latitude, but still on the fringes of the northern Sudanian climate (as defined by Peron and Zalacain, 1975), *D. rotundata* yams are still frequently grown in the soils of hydromorphic bottomlands, notably along the Burkina Faso-Ghana border and in the Pama region of Burkina Faso and Benin. The same form of agriculture is also encountered in Benin (on the plains adjacent to the Atakora Range), Togo and Mali. All of these yam-growing activities in refuge biotopes suggest that climatic drying has gradually marginalized the growing of *D. rotundata* yams in a territory where it was once much more widespread. This idea is sometimes supported by the collective memory, for instance in the Banikoara region of Benin (Dumont, personal observation).

Anthropogenic pressure

African yams have probably been a source of food for humans since time immemorial and were very likely also eaten by their hominid ancestors. This idea, first advanced by C.M. Hladik in 1985, was taken up again by Dounias in 1996. More generally, in the words of Brunet and Picq (2001), "It can safely be affirmed that the [African species of] Australopithecus used sticks to dig up the underground parts of plants [ ] and perhaps used quartz splinters to cut or scrape at the vegetation." This takes us back more than 2.5 million years into the past. It can be supposed that the wild yams of Africa have been manipulated since the dawn of time, with varying subsequent degrees of genetic drift. The most advanced results of this evolution would be the development of the Dp2 form of *D. praehensilis* and the *D. abyssinica* yams of West Africa.

Coursey (1976) suggested that a rudimentary form of yam cropping might have come into being at the end of the Pleistocene period (35,000 BP)[1], when the climatic drying of West Africa undermined food security based on foraging. The development of the lithic industry (before 10,000 BP) followed by the advent of metal tools (around 4,500 BP) would have led to successive improvements in the way yams were utilized. Coursey also indicated that several early European explorers of West Africa reported that yams were one of the crops cultivated in the region, and there is documentary evidence of a yam trade in Nigeria dating back to 1505.

In many cases, foraging seems to have been the first step towards domestication. Bahuchet *et al.* (1991) noted that this was a logical sequence of events that started with gatherers accumulating technical information about the plant material. Bahuchet (1982)

---

1. BP: before present (reference year: 1952).

also uses the term 'semi-cultivation' when describing the systematic replanting of tuber heads by Aka hunter-gatherers of the Central African Republic. This practice, also observed in southwestern Ethiopia (Hildebrand, 2003), exists in forested regions of Guinea (Dumont, personal observation) and in the southern part of Benin (Mignouna and Dansi, 2002), involving wild yams of the Enantiophyllum section. Dounias (1996) also documented it in eastern Cameroon, and stressed that each wild yam plant treated in this way is appropriated. The wild plant material is managed in its natural environment and a so-called paracultivation system is introduced (Dounias, 2001), thus reducing the random nature of the foraging and involving selection of the natural diversity. The selection criteria common to all these situations are probably the useful tuber volume, organoleptic qualities and ease of uprooting. It can be supposed that this selection pressure models the variability of wild yam populations.

Another system of wild yam appropriation observed in forested regions of Africa was regarded by Chevalier (1936) as a form of protocultivation. In this case, the plant material is removed from its natural biotope and either planted in natural fallows in the vicinity of the village or (more recently) combined with moderate shade-tolerant cash crops (cocoa, coffee). These yams retain their morphological traits, including exuberant growth of the vegetative organs. For the latter reason, they have to be planted next to tall (often live) stakes. The plants can then remain in place for several decades, giving one early or late harvest per year.

The Dp2 morphotype of *D. praehensilis* is a yam of the Enantiophyllum section that is used in many protocultivation situations. Kaape bolonda yams of western Guinea, Cocoassie yams of southern Côte d'Ivoire and Gban yams of southern Benin are in this category. In Cameroon, the Ngoro and Tikar people collect *D. praehensilis* yams in the forest and plant them in their home gardens, while the Doupa and Dourou do this with *D. abyssinica* savannah yams (Dumont *et al.*, 1994). Cultivated *D. praehensilis* yams were encountered by Hladik *et al.* (1984) in forest areas of the Central African Republic and Gabon. These yams have not been markedly modified by human manipulation, but they perform to expectation when integrated in a shifting mixed cropping system with a ready supply of natural stakes. For this reason, and also because cassava and plantain are the main food crops locally, it has not been necessary to exert high domestication pressure on yams from African forest regions that are used in vegeculture.

Protocultivation can, when necessary, considerably advance the production period. From *D. praehensilis* yams, farmers obtain plant material that can be used for double harvesting. It is thought that they achieve this by selecting from the diversity of the Dp2 morphotype. As already mentioned, this wild population is subject to early and late foraging in the westernmost part of the African forest region. Farmers in the Nigerian state of Benue also distinguish between early- and late-maturing wild yams, which would be the Dp2 form of *D. praehensilis* (Vernier *et al.*, 2003). Protocultivation-induced selection creates a cultigen population whose production earliness is a trait that is also typical of *D. abyssinica*. This common trait can be added to the previously mentioned morphological resemblances between the two types of yam, thus further supporting the hypothesis of an evolutionary phenomenon that places *D. praehensilis* and *D. abyssinica* in the same genetic group. As early as 1898, Baker wrote that "these two yams are related and perhaps not fundamentally different." More recently, Hamon (1987), Terauchi *et al.* (1992), Ramser *et al.* (1997) and Chaïr *et al.* (2005) did not separate *D. praehensilis* from

D. abyssinica. Lastly, the study of Scarcelli (2002) highlighted cases of close genetic proximity between these two wild yams in both northern and southern Benin.

Protocultivation is a system that can ensure a food supply in the rainy season when foraging is more difficult, or even more haphazard when there is a scarcity of early Dp2 yams in the wild flora. On three occasions, Hamon (1987) identified a zymotype common to two Cocoassie yams found in protocultivation conditions in forested regions of Côte d'Ivoire, thus highlighting that genotypes derived from the wild population had been propagated vegetatively. Dounias (2001) pointed out that vegetative propagation begins when paracultivation is implemented. Coursey (1976) claimed that protocultivation is part of the process leading to yam domestication. This is debatable, however, with respect to humid regions. The studies of Hamon (1987), Mignouna *et al.* (1998) and Scarcelli (2002) showed that the genetics of Cocoassie, Gban and Kaape bolonda cvs used in protocultivation in Côte d'Ivoire, Benin and Guinea, respectively, were not very close to *D. rotundata*—they appeared to be (Dp2-type) *D. praehensilis* yams with morphological traits that can only be partially modified by human manipulation. Hence they seem to be cultivated *D. praehensilis* yams rather than *D. rotundata* yams. They would be equivalent to *D. praehensilis* yams used in vegeculture systems of forest areas of Central Africa (Hladik *et al.*, 1984). The results of Assogba (1993) indicated that they are surviving vestiges of a very ancient form of West African agriculture, probably practiced when the forest ecosystem predominated in areas now transformed into savannah.

Coursey (1976) was probably correct in pointing out that the African Enantiophyllum species might have been domesticated along the entire length of the forest/savannah interface that crosses West Africa from east to west. In this ecotone, wild yams are still often cropped (Hamon, 1987). Protocultivation—perhaps a less rudimentary form of agriculture—would have started with forest yams, as suggested by the relictual nature of various hexa- or octoploid cvs in the savannahs of West Africa, which are probably related to the *D. burkilliana* yams growing in degrading forests. Domesticated cvs have also been created from *D. praehensilis*. As already noted, both processes are still implemented in African forest regions.

However, it is in savannah areas where *D. rotundata* yams predominated in agriculture. Under environmental conditions unfavorable for ensuring food security, domestication diversified the plant material according to two complementary objectives, i.e. extending the production period and the storage life of the crop. The wild yam species initially used as the source of plant material was likely *D. lecardii* (still domesticated in Cameroon) and *D. praehensilis*, which can still be found in the remaining vestiges of mesophyll forest or its savannah ecotypes.

It is not known exactly where *D. rotundata* yams were first domesticated for cropping in savannah agriculture conditions. From Côte d'Ivoire to eastern Nigeria, behavioral patterns associated with these yams are similar in all the ethnic groups involved. The cultivation of *D. rotundata* yams is a strictly male prerogative. They are traditionally eaten in the form of futu. It is the only crop associated with ritual practices, which are relatively similar everywhere (Coursey, 1976). It is thus difficult to determine the primary locus of domestication, although Coursey (1976) pinpointed it in a preforest zone extending across Nigeria and Benin.

The area of *D. rotundata* domestication seems to have spread centrifugally to the savannah zone. Different arguments suggesting that domestication is a recent phenomenon in

Côte d'Ivoire and Guinea have already been mentioned. The entire western boundary of the African continent, including Guinea, is now affected by this phenomenon.

The situation differs in eastern Africa. The *D. rotundata* plant material used in West Africa is found only in part of the Cameroonian savannah (Hamon, 1987; Dumont *et al.*, 1994), with perhaps some minor spillover in neighboring Chad. This is probably due to the long troubled period that Cameroon has gone through in the past. From the end of the 18th century, inhabitants living north of the Adamaoua region took refuge in the highlands (Seignobos, 1998). Very little arable land was available there, so *D. rotundata* yams could not be grown and were lost (apart from a few exceptions). One such exception was cv Kokou of the Bamileke living in the Foumbot region. This yam has been recorded under twelve other vernacular names in western and northern Cameroon (Dumont *et al.*, 1994). It was grown in earth-filled faults in the rock and probably owes its survival to its high ritual value. It is cultivated in pits and produces the large tubers that are essential for social relations, although some of these tubers are now sold commercially.

It was only at the beginning of the 20th century that the Cameroonian people who had fled to the mountains gradually returned to the plains to continue their domestication of wild Enantiophyllum yams. However, current commercial production of *D. rotundata* yams owes little to these activities. The original cv Bakokae, which is certainly very old but of debatable origin, was initially cropped. Now an increasing number of Nigerian cvs are being introduced and, in some parts of the Dourou region, they account for 30% of the total yam cultivation area (Seignobos, 1997). This use of foreign plant material indicates that domestication, which was resumed at a very low level, is currently not very efficient. The long period of social tension in the history of Cameroon has apparently nullified earlier domestication achievements, thus stalling its eastward spread. This idea was first put forward by Hamon in 1987.

Reciprocal gene flow between wild and cultivated yams

It is suspected that reciprocal gene flows have occurred between sexually fertile *D. rotundata* yams and their wild relatives, based on the following arguments. As discussed later, domestication involves collecting tubers from sexually functional wild plants and subsequently cropping them. The genotypes are not modified so there is no barrier to sexual reproduction. Beninese farmers often notice certain morphological traits of their local *D. rotundata* yams in *D. abyssinica* biodiversity (Baco, 2000; Okry, 2000). Seventeen of the 32 clones of Beninese *D. abyssinica* yams in the course of domestication studied by Scarcelli (2002) appeared to be genetically very close to *D. rotundata*.

The traditional organization of Bariba agriculture promotes reciprocal gene flow between wild yams and their domesticated forms. Cultural prohibitions have long prevented the growing of yams in a concentrated area (a yam plot should not be within sight of any other surrounding plots). Even today, over 50% of Bariba farmers surveyed still preferred to isolate their yam plots (Baco, 2000). This clearly facilitates genetic mixing between wild yams (*D. abyssinica* and *D. praehensilis*) and sexually functional early *D. rotundata* cvs. The latter are few in number and represent less than 50% of the total cultivated plant material. A survey of domestication practices in northern Benin by

Dumont and Vernier (1997a) recorded only seven of them (38.9% of the overall number), some with a very limited geographical distribution. However, each cv is extensively copied by vegetative propagation. Between 60 and 200 copies per hectare of yam crops can be found (Dumont, 1997) and the process is repeated annually in the area cultivated by each village. In many locations, sexually functional cultivated genotypes predominate in terms of the overall supply of gametes or zygotes, as compared to the wild yam population, which is readily genetically modifiable due to their chronically low numbers. The reasons for these low numbers will be discussed later.

This reproductive advantage enhances the genetic proximity between cultivated and wild yams and suggests that *D. lecardii* and *D. praehensilis* could have been initial partners in a coevolution process. *D. abyssinica* and *D. rotundata* yams might be late products of this coevolution system. In savannah areas, *D. abyssinica* is nowadays domesticated to obtain *D. rotundata* cvs.

The long interruption in the coevolution process in northern Cameroon probably explains the current lack of achievement in yam domestication, which has only recently been resumed in this region. This highlights that other environmental changes such as excessive clearance of savannah woodland, and the detrimental effect of trade on cultivated yam diversity, might endanger the traditional domestication process. The latter is still the most productive source of genetic innovation with respect to *D. rotundata* yams.

This coevolution process *a priori* concerns only double-harvest *D. rotundata* yams that are still sexually functional. However, it probably also has an impact on late cvs, as they appear to originate at a later stage than phenomena that give rise to early yams. There is very little diversity among late *D. rotundata* yams from Côte d'Ivoire, where domestication is a relatively recent practice, which also explains the success of *D. alata* yams there. More generally, nowhere in Africa has agriculture relied solely on late yams, at least until recently. In contrast, early *D. rotundata* cvs are often used in traditional agriculture, either alone or in combination with their late-maturing equivalents.

As previously noted, scientific research has not been able to separate *D. praehensilis*, *D. lecardii*, *D. abyssinica* and *D. rotundata*. It can therefore be supposed that all four yams belong to the same gene pool, which means that interbreeding should be possible. Indeed, on the basis of specific morphological criteria in addition to floral sterility (partially demonstrated), late *D. rotundata* yams might be interspecific hybrids. This interpretation, as already noted above, is discussed in detail later.

Effects of periodic fallows

In traditional agriculture, which until recently has been the standard strategy, *D. rotundata* yams are grown as a break crop after clearance of wooded savannah—usually the result of fallow regrowth. Traditionally, for the Bariba people of northern Benin, the end of the fallow period used to coincide with a new generation of farmers, i.e. with each cycle lasting between 20 and 30 years. The fallow period is now often shorter, however, because of increased population growth and sometimes the fact that commercial crop plots are concentrated along the roadways. The following information therefore applies to a situation that virtually no longer exists but which was probably crucial in the domestication of *D. rotundata* yams currently being grown.

As already discussed, sexual propagation of *D. praehensilis* and *D. abyssinica* yams is highly favored by disturbances within their natural ecosystem. Agriculture is certainly the most important disturbance factor. In its dynamic phase, the reforestation of fallow land creates an environment in which yams can prolifically multiply from natural seedlings. The protection provided against sunlight and soil crusting due to rainfall facilitates the germination of windborne seeds from the wild and crop fields. A large supply of natural staking is available for the new seedlings if the woody regrowth remains dense. Moreover, the shade provided by this cover prevents the growth of herbaceous annual vegetation, thus reducing damage from annual fires. Under these conditions, *D. abyssinica* and *D. praehensilis* yam populations can reach several thousand plants per hectare (Dumont, 1988).

For two main reasons, wild yam populations decline after this intense multiplication period. Firstly, temporary waterlogging has a negative impact, which increases as the yams grow older since the tubers have a greater probability of encountering water rising from the water table as they grow in length. Secondly, scrub vegetation that can serve as natural stakes for yam plants becomes much scarcer as the reforestation process comes to an end.

The wild yam population falls to a very low level under this double pressure. Less than five *D. abyssinica* plants per hectare of long-term fallows were sometimes observed by Dumont (1988) in northern Benin, while Hladik *et al.* (1984) recorded a maximum of 10 *D. praehensilis* plants per hectare in the forest area of the Central African Republic. This low yam population level has two consequences. Panmixia is no longer always possible within the fallows. In addition, crosses frequently occur between successive generations as the progenies can survive only in refuges, such as fossil termite mounds and rocky areas already colonized by the parents. On the other hand, because of their lack of number, the remaining wild yams have a high probability of acquiring genes from sexually active *D. rotundata* yams. As just discussed, the latter are abundantly represented in current agriculture but their genetic base is relatively narrow.

In short, periodic fallows have two successive effects. Initially, genetic variability is reestablished and probably reorganized, as the probabilities of recombination and segregation have increased considerably. This enhances the genetic and phenotype diversity of wild yams while occasionally changing their homeostatic equilibrium. Secondly, the allelic diversity of the wild yam population is reduced due to consanguinity induced by the decline in wild yam numbers and by gene flows from *D. rotundata* yams.

Each fallow period thus upsets the genetic equilibrium of wild yam populations, and then a new one is established on the basis of the environmental changes that occur. These repeated adaptations constitute what we have called the coevolution process. This would explain why the best domestication results have been obtained in regions that have been subjected to high long-term anthropogenic pressure. There are two reasons why this pressure is the real driving force behind the coevolution process. Firstly, more farmers are involved in domestication, thus increasing the efficacy of selection among wild *D. praehensilis* and *D. abyssinica* yams. Secondly, many farmers cultivate large areas of *D. rotundata* yams, therefore generating high gene flows to the wild parents—this gradually increases the average value of these parents in terms of suitability for domestication.

Variations in genome expression

The phenotype is the result of the interaction between the genotype and the environment. A change in the latter can lead to profound modifications in the phenotype. A genome can thus have various phenotypes. Epigeneticism is the underlying phenomenon here since, as far as we presently know, new traits obtained in this way by *D. rotundata* yams are transmissible only by vegetative propagation.

Transferring wild yams to a cropping environment subjects them to a drastic change of ecosystem and specific cultivation techniques. According to Chikwendu and Okezie (1989), these artificial pressures are responsible for major morphological transformations in *D. praehensilis* yams—the vegetative organs become much smaller, the tuber shape less irregular and, most importantly, the fiber content of the tuber decreases and its starch content increases.

Sporadic or short-term morphological variations were previously reported in *D. rotundata* yams. Trilobed leaves have only been noted once (northern Benin) in 14 years in cv Boni Oure (Dumont, personal observation). In the same cv, stem fasciation is more frequent but always limited to the germination period. These events highlight that ontogenetic disturbances can be manifested in different ways.

Ambivalent ontogenetic functioning is regularly induced by double harvesting. This has one key effect. Between the first and second harvest, the same plant, i.e. the same genome, expresses the morphological tuber traits differently within a single crop season.

Sustained deregulation of this process, rather than interspecific hybridization, might be an alternative explanation for the appearance of kokoro-type *D. rotundata* yams. As already noted, their natural tuberization resembles that occurring during the second harvest in early-maturing cvs. This behavior—which previously had to be induced through a particular cultivation technique—has become spontaneous. The genome would thus have switched from its primary phenotype expression and adopted another one.

If kokoro yams are the result of the process described above, overcropping of the land might induce their genesis. Indeed, kokoro cvs are more numerous among ethnic groups that put heavy pressure on the land (Kabye in Togo, Yom and Nago in Benin, Yoruba in Nigeria). These are the only yams that give a satisfactory yield on degraded soils, which might account for the large number of kokoro cvs cultivated by the ethnic groups concerned, but it fails to explain the rich diversity of the local yam plant material.

Nevertheless, the shift from double-harvest to kokoro-type *D. rotundata* yams would be a gradual process with an intermediate stage, i.e. mixed cvs. In addition, the process would depend on long-term human pressure. Kokoro cvs thus seem to be an incidental product of a very old form of yam cultivation. The primary center of domestication of *D. rotundata* yams would therefore be a geographical area that includes areas in which kokoro yams are abundant, i.e. Western Nigeria, Benin and Togo. This brings us back to the idea put forward by Coursey (1976), but slightly refocused.

Mutation

Several authors (Miège, 1952; Ayensu and Coursey, 1973; Hamon, 1987; Dansi *et al.*, 1999) consider that somatic mutation has been involved in yam domestication and diversification. This hypothesis has not yet been verified experimentally. In principle, morphological or other transformations determined by somatic mutation are not, in the first generation, aligned with a corresponding genotype modification and are therefore not transferable through a coevolution process. Coevolution thus only has a marginal role in the process leading to double-harvest yams, but this needs to be qualified. As there is no clear separation between germen and soma in plants (Hallé, 1999), any mutation in the latter can be passed on to the former. The fact remains that somatic mutations are erratic, although perhaps not rare on a large population scale. Many, however, remain silent while others can have an adverse or, on the contrary, beneficial effect in agriculture.

On the other hand, allelic mutation and chromosome mutation directly initiate sexually inherited phenotype variations. A mutant may gradually take over within its environment if it has a higher adaptive value than the rest of the yam population. Yam domestication fosters this elitist dynamic process. African farmers gather, select and (vegetatively) propagate mutants that will introduce improvements or even innovations in their plant resources. Since they are hereditary, mutations promoted in this way can be integrated in the coevolutionary system that is assumed to link sexually functional *D. rotundata* yams to their wild parents. Mutation thus appears to be a potentially powerful tool for diversification, or even evolution, provided that it is not lethal or deleterious. However, according to Valdeyron (1961), mutation has a limited role in the genetic transformation of cultivated plant material: "Mutation seems to be far less important than migration. The latter, and zygotic migration in particular, is the domestication pressure par excellence that seems to have conditioned the transformation of the plant material used in agriculture today, as human exchanges have probably played an essential role in the spread of cultivated plants." Concerning *D. rotundata* yams of West Africa, as already noted, zygotic migration has been facilitated by the retreat of the mesophyll forest. All of these ideas relating to genetic mutations clearly only apply at the micro-evolutionary level.

Ploidization

Domestication does not appear to take advantage of any ploidization phenomena and there is nothing to suggest the presence of such phenomena during hybridization between cvs or with the wild parents. In addition, there is a clear absence of correlation between agronomic advantage and ploidization with respect to yams of African origin belonging to the Enantiophyllum section. Yam production in Africa now mainly involves *D. rotundata* yams, all of which are tetraploid like their wild ancestors. However, cultivated octoploid or hexaploid yams (*D. cayenensis* and satellite forms) play a very marginal role in West African agriculture and have not been affected by technical changes in yam production as a result of emerging socioeconomic constraints. Mention could also be made of the edible wild species *D. minutiflora* Engl., which has never been domesticated although, according to Miège (1954 and 1958), its ploidy level (2n = 120) is one of the highest among the Dioscoreaceae. Lastly, concerning Enantiophyllum yams of Asian origin, it can be noted that *D. alata* production in Côte d'Ivoire is largely based on the Florido and Bete-Bete cvs, both of which are tetraploid (Gamiette *et al.*, 1999), while the hexaploid cvs introduced for research purposes (De Agua and Feo) have proved to be either unacceptable for culinary use or low yielding (Dumont, 1994).

Domestication leading to *Dioscorea rotundata*

Definitions and general aspects

Domestication is a set of practices that are applied to edible (or potentially edible) wild plants to facilitate and sustain their food use. In more technical terms, it involves the adaptation of wild plant material to the environmental conditions of agriculture. This is achieved by changing the genetic equilibrium of the initial populations and enhancing their traits through genetic innovations (Kupzow, 1980).

The starting material for the domestication of *D. rotundata* yams is a population of wild morphotypes that have well-developed vegetative organs, sexual vigor, and a small and relatively bitter tuber that is difficult to harvest because it is often long and sometimes branched. Domestication attenuates or eradicates these characteristics, which farmers view unfavorably.

The aerial architecture of the plant is substantially modified during the domestication process. Stem internodes become much shorter, with a concomitant equivalent reduction in the stem length. Ramification occurs on the lower part of the stem. There are not many primary branches (as in the wild) but each develops a large number of secondary branches that are quickly covered with thick foliage. These various transformations condense the mass of the aerial vegetative organs, thus reducing or eliminating the need for staking.

The tuber is also enhanced. It becomes shorter and thicker, with fewer epidermal roots, and the flesh takes on a uniform light color. While controlling these phenotype modifications, the farmer also selects according to the physiological traits, i.e. production period, storage life, seed potential (in the different forms), and sometimes the ability to produce a large number of tubers. More general criteria relating to the plant's biological plasticity and yield potential are also taken into account for selection. However, domestication is mainly geared towards improving cooking quality. Various criteria linked with this trait were discussed earlier. The procedures used by African farmers to obtain a satisfactory domestication result will be examined later.

All edible wild yams in Africa appear to have long been subject to domestication pressures, with extremely varied results. *D. semperflorens* has withstood all attempts at domestication and *D. schimperiana* (a taxonomically close species) has not reacted much better. Other species have responded more positively and become important in local agriculture. Two *D. mangenotiana* cvs (Samancou and Jagana) are known to exist, one in Côte d'Ivoire (Hamon, 1987) and the other in Cameroon (Dumont *et al.*, 1994). The *D. bulbifera* species has been domesticated in both West and Central Africa but is almost exclusively grown in western Burkina Faso and primarily in the Bamileke region of Cameroon. In the latter region, the *D. dumetorum* species has been subject to high domestication pressure, resulting in several yellow-fleshed cvs of excellent cooking quality (Lyonga, 1976; Trèche, 1989). The domestication of *D. burkilliana* was likely even more productive. Two distinct results would be possible, as already indicated. One would be obtained by hybridizing *D. burkilliana* with a wild species whose identity is still just speculated (Hamon, 1987; Terauchi *et al.*, 1992; Mignouna *et al.*, 1998) while the other would be a polyploid series (like *D. alata*) developed by *D. burkilliana* (Dansi *et al.*, 2001). However, yams with the greatest vegetative diversity and widest use in African agriculture and trade are the result of domestication of species that are presently identified as *D. abyssinica* and *D. praehensilis*.

Rationales and limits of domestication

Significance and importance of domestication for farmers

Very few surveys have involved interviews with African farmers to determine their reasons for domesticating wild yams. However, the explanations that have been recorded often mention episodic threats to the productivity of *D. rotundata* yams from occult powers, so it is periodically necessary to replace cultivated yams by others collected in the wild. The results of three recent surveys in Benin (Baco, 2000; Okry, 2000; Vernier *et al.*, 2003) more clearly identified the motivations underlying domestication. These can be condensed into three basic ideas, which are presented below in random order.

Wild yam domestication techniques are bequeathed by the ancestors and should be put into practice and transmitted to the younger generations. Baco (2000) reported that over half of the farmers domesticating yams in northern Benin who were interviewed in his survey were under 40 years of age, so the know-how had clearly been passed on.

The regular planting of wild yams amongst cultivated cvs introduces 'new blood', thus making the latter more agronomically productive overall and improving their cooking quality.

The domestication of wild yams enables farmers to develop a pool of plant material that can be used to start, develop or diversify their yam production. Note that for a long time African farmers organized food security on the basis of a self-sufficiency strategy. In such conditions, it was mainly only possible to obtain seed tubers through traditional procedures governing plant resources. These constraints also apply to migrant farmer populations that are currently settling in the yam belt, but they are now becoming easier to overcome through different forms of exchange.

It should also be stressed that the collective memory of savannah peoples often includes periods of severe famine during which it was necessary to eat the seed tubers. Following these periods, wild yams were used to kickstart yam cultivation.

In the light of these ideas, present-day farmers clearly still domesticate yams just as their ancestors did for centuries. As yet there appears to be no discontinuity between the present and the past, so this is a vast potential field of ethnobotanical investigation. Note that rural societies in Africa today have a negative perception of domestication, which is considered as a demeaning activity, and farmers practicing domestication are regarded as being unable to produce sufficient yams to feed their families. Only notoriously wealthy farmers can acknowledge that they domesticate yams without being socially denigrated (Baco, 2000).

Another factor that currently makes domestication less interesting is the fact that plant material can now be easily obtained from sometimes faraway yam-producing regions. Many *D. rotundata* yams currently grown in Benin were recently introduced from Nigeria and Ghana following human migrations. Yams have thus also been introduced in various African regions from other cropping areas. Plant material from eastern Nigeria is now widely used in the English-speaking part of Cameroon, while cvs specific to northern Nigeria are currently cultivated in Chad and the Central African Republic (Pfeiffer, 1987).

Provided that its dissemination is properly organized, the biodiversity of *D. rotundata* yams should be able to at least temporarily solve many of the occasional problems now arising with yam production in Africa and elsewhere (Dumont and Vernier, 1997b). Farmers are still convinced that it is essential to regularly renew the plant material. *D. rotundata* yams have a temporally limited existence—inhabitants of most African villages talk about yam cvs that have disappeared. This finite existence of *D. rotundata* yams is a natural phenomenon and sometimes taken into account in traditional practices. Among the Dourou of northern Cameroon, for instance, declining cvs are replanted in the village to finish their existence there (Dumont *et al.*, 1994).

The reasons for the natural disappearance of yam cvs are not known. Farmers explain this phenomenon on the basis of a decline in yield or an inability to adapt to climatic changes. One could suspect an accumulation of deleterious mutations or pathogens (notably viruses) in the plant. Another possible reason is an inability to adapt to environmental changes. This inability might be the result of repeated vegetative propagation over a very long period of time, as cvs that have lost their sexual reproduction capacity, and are therefore disconnected from domestication, cannot adjust their homeostatic equilibrium.

It is likewise virtually impossible to determine how long a particular yam cv will last. There are only two historical benchmarks, i.e. the introduction of cv Kponan in cropping systems in Côte d'Ivoire and the transfer of numerous African yams to the West Indies and South America. Both events took place over 200 years ago, whereas the yams concerned come from germ lines originating even further back in time.

Successive studies undertaken in Benin and Nigeria (Dumont and Vernier, 2000; Vernier *et al.*, 2003) revealed that 40 to 90% of farmers interviewed know how to obtain *D. rotundata* yams from wild yams (now identified as *D. abyssinica* and *D. praehensilis*). However, this knowledge currently appears to be put into practice on 3–20% of the farms surveyed. In two studies conducted in Benin, Baco (2000) and Okry (2000) reported that 14 and 16.7%, respectively, of yam growers were domesticating yams.

Our view of what domestication represents in terms of agriculturally useful results is based solely on data collected in Benin. In a national survey, Vernier and Dansi (2000) identified 36 *D. rotundata* cvs that farmers said had been domesticated in the recent past, but 22 of these were found among the Yom and Nago peoples, who put heavy pressure on the land. The cultivated plant material has to be replaced when agricultural conditions become difficult. There are two ways of overcoming the problem other than by introducing new yams. One is to select cvs via domestication that are adapted to the environmental changes that have occurred. The other is to promote cvs that are proven to be hardy, but which have not been extensively used in production conditions (e.g. Alassora yams of the Kabye region of Togo, which are related to kokoro yams). The resulting revitalization of the plant material is often wrongly interpreted as being due to domestication. Allegedly recent cvs can actually be very old. This situation is very likely with plant materials that have a low sexual reproduction capacity.

Dumont and Vernier (1997a) extrapolated that, at the time of their survey, 1,600 domestication attempts were under way in the Bariba region of northern Benin (in a 10,000 km^2 area). Activity on this scale should have produced a large number of innovations each year, but this was not the case. Local agriculture is based on the use of about 40 cvs that until recently seemed to be relatively stable. According to collective memory, much of this plant material comes from earlier local domestications. It is also known that nine cvs are of foreign origin, but in several cases they could have been adopted by the Bariba peoples a very long time ago. Two examples of this are the (single-harvest) Gambari Gninou and Yon Bouanri cvs, both linked to precolonial history. One appears to have been brought by the caravan trade that passed through northern Benin until the 19th century, linking the Hausa region of Nigeria with the Ashanti kingdom of Ghana (Lombard, 1965). Legend has it that the other was offered as a gift at an aristocratic marriage celebrated over eight human generations ago.

The low number of cultivated cvs, the genetic age of the plant material used and the way that the introduction of new yams is viewed as a happy event all suggest that domestication generates few new cvs. However, as noted later, this is often not what the farmers are aiming at. Domestication creates a large amount of genetic diversity, which usually remains within existing cvs and is thus often inconspicuous. It relates to characteristics such as adaptation to the environment, productivity or cooking quality (especially the suitability for futu-making).

Framework of domestication

It is difficult to determine the respective contributions of *D. praehensilis* and *D. abyssinica* to the domestication process leading to *D. rotundata* yams. Dansi *et al.* (1999) reported that, based on morphological criteria, eight of the 23 *D. rotundata* groups identified in Benin were linked to *D. praehensilis* and nine to *D. abyssinica* while the remainder combined the morphological traits of both species. Using RFLP analysis, Scarcelli *et al.* (2005) revealed genetic links between nearly 50% of the clones that are being domesticated in Benin and the *D. rotundata* cvs already grown there. *D. abyssinica* plays a major role in the other cases. It would be risky to apply these ratios to the whole of West Africa. Several hypotheses were previously put forward which suggest that virtually all domestications start from *D. praehensilis*, particularly in Côte d'Ivoire.

Collected information indicates that domestication rarely produces single-harvest *D. rotundata* yams. They are included in a general domestication strategy, but farmers do not control their creation, they simply collect morphotypes that have been transformed without their intervention. Two hypotheses have been put forward to explain this natural transformation:

– Dansi *et al.* (1999) suggested that single-harvest cvs are interspecific hybrids between *D. praehensilis* and *D. abyssinica*. Is this still valid, given that several arguments suggest that both wild yams belong to a single species that has been morphologically fragmented by climatic change and remodeled by anthropogenic pressure?

– Late-maturing cvs have traits that are independent of *D. praehensilis* and *D. abyssinica*. This particularly applies to the tuberization model of kokoro yams *stricto sensu*. We previously tried to explain this divergence on the basis of a 'neo-Lamarckian' theory that assigns an important role to the environment in gene expression. It could be assumed that this phenomenon affected all *D. rotundata* yams, with the transformation of double-harvest cvs into mixed yam cvs and then, depending on the prevailing agricultural constraints, into large-tuber late-maturing yams or the kokoro morphotype. Here we pooled various information from our study in support of this idea. Domestication only provides double-harvest *D. rotundata* yams. Late-maturing kokoro yams have never been found in the wild flora. There are few or no late-maturing (especially kokoro) yams in agricultural systems with a recent domestication history (Guinea, Côte d'Ivoire, Cameroon). Floral sterility—which is apparently common in late-maturing yams—has been noted in some mixed cvs (Krenglé in Côte d'Ivoire, Boni Ouré in northern Benin), but has never been detected in early-maturing yams.

A third hypothesis can be added to the two already mentioned. As there are intermediate forms between *D. praehensilis* and *D. abyssinica* in savannah flora, it is likely that these are the origin of non-kokoro late-maturing *D. rotundata* yams. This category of yams would be *a priori* fertile and include female components. All currently known female late-maturing yam cvs are sterile.

If the sterility of single-harvest *D. rotundata* yams is confirmed, this will indicate that one form of domestication harnesses genetic combinations that are incapable of perpetuating themselves in the wild by sexual reproduction. Plant material created in this way could thus only be maintained in agricultural conditions by vegetative propagation. As noted earlier, the *D. cayenensis* yam is another African construct of the same nature and for which only male plants are known. There would thus be two homologous phenomena, one occurring in savannah areas and the other in forests, and both are probably linked to the same evolutionary dynamics.

Domestication by farmers produces only double-harvest *D. rotundata* cvs. This was clearly established on the basis of the results of a first survey in Benin (Dumont and Vernier, 1997a) and fully confirmed by later studies conducted in the same country (Baco, 2000; Okry, 2000). Observations in eastern Nigeria (Vernier *et al.*, 2003) led to the same conclusion. The results of the initial survey also showed that domestication is organized only around cvs that are still sexually functional, which is still in line with the supposed coevolution process.

As already discussed, present-day domestication initiatives leading to double-harvest yams utilize only part of the available wild plant material, i.e. the Da2 morphotype for *D. abyssinica* and the Dp2 morphotype for *D. praehensilis*, with both more likely to have the desired traits. This situation is probably largely determined by anthropogenic pressures, with certain genotypes from the wild population being favored over others. This selective process begins with paracultivation and becomes more effective with protocultivation, whereby advantageous genotypes are propagated vegetatively. It has been underlined how this selective approach can easily modify the genetic equilibrium of wild yam populations and gradually give them cultigen features, thus genetically linking them to cultivated yams.

Historically, the yam domestication front in West Africa is a shifting zone that follows the southward or westward retreat of the mesophyll forest. Furthermore, there is an indication that the only species initially domesticated in areas now covered by savannah were *D. praehensilis* and/or its supposed *D. lecardii* ecotype, which is still domesticated in Cameroon. *D. praehensilis* thus emerges as the pioneer of both *D. rotundata* yams and the apparently commensal wild *D. abyssinica* yams of West Africa.

The assumed coevolution of cultivated yams and their wild parents can be in different states of advancement since African yams have been domesticated for varying lengths of time. Schematically, however, there appear to be two quite distinct situations.

At the forest ecotone level, *D. praehensilis* yams would be influenced by the centrifugal effect of a more sophisticated (because older) domestication process practiced in the neighboring savannah area. Such gene migration represents an evolutionary pressure that can increase allelic homogeneity within a set of populations (Gouyo *et al.*, 1997), provided of course that there is a relatively high degree of genetic compatibility. These ideas are very close to those of Valdeyron (1961) referred to in the chapter dealing with the possible role of mutations in the transformation of plant material. For the present purposes, the key point is that the sympatry of *D. praehensilis* and *D. rotundata*, as facilitated by degradation of the mesophyll forest, would enhance the potential appearance of new *D. rotundata* cvs. Mignouna and Dansi (2002) thus reported the recent creation of three double-harvest cvs in southern Benin, while Dumont and Vernier (1997a) encountered only one case in the northern region, even though a high number of domesticating farmers were interviewed.

At least in its primary stage, the first domestication model discussed often results in unsatisfactory products. The tubers may be branched or digitated, their overall shape is sometimes unstable, and the upper part of the tuber extremely fibrous and therefore inedible. Cooking quality can also be a problem. This aspect could be improved, particularly in terms of suitability for futu-making, through gradual adjustment of the genetic diversity by repeated domestications in savannah conditions. Furthermore, it should be noted that yams are generally not converted into futu in forest regions of Africa.

In the oldest domestication area, i.e. the savannah, the coevolution process would have transformed Da2 *D. abyssinica* yams into a highly cultigen population closely linked to the *D. rotundata* yams that are still sexually functional there. This process would function differently, with a cumulative effect for each of the partners concerned, and resemble recurrent selection. Genetic combinations selected as phenotype traits in the wild plant material would be reproduced in large quantities by vegetative propagation in agricultural conditions, and then be sent back to wild yams. Domestication would take advantage these genetic combinations, which are then modified by sexual reproduction in the wild, sometimes occasionally giving rise to improvements of agricultural interest. The *D. rotundata* population would play a decisive role in the coevolution process when it becomes numerically dominant. Sexual reproduction among wild *D. abyssinica* and *D. praehensilis* yams would then function within the genetic limits of their corresponding cultivated forms, thus considerably reducing their range of natural variability. In other words, *D. rotundata* cvs would be very genetically close to their wild parents when domestication begins (as is the case with cv Lokpa in Côte d'Ivoire and cv Gnidou in Benin), but then the situation would be reversed. Once this stage is reached, it is unlikely that domestication will to lead to major innovations, but it does generate small-scale phenotype modifications that are occasionally beneficial to recently domesticated cvs, and it sometimes even gives rise to variants. According to Dumont and Vernier (1997a), these are the two directions followed by present-day domestication in the savannah region of northern Benin (figure 5).

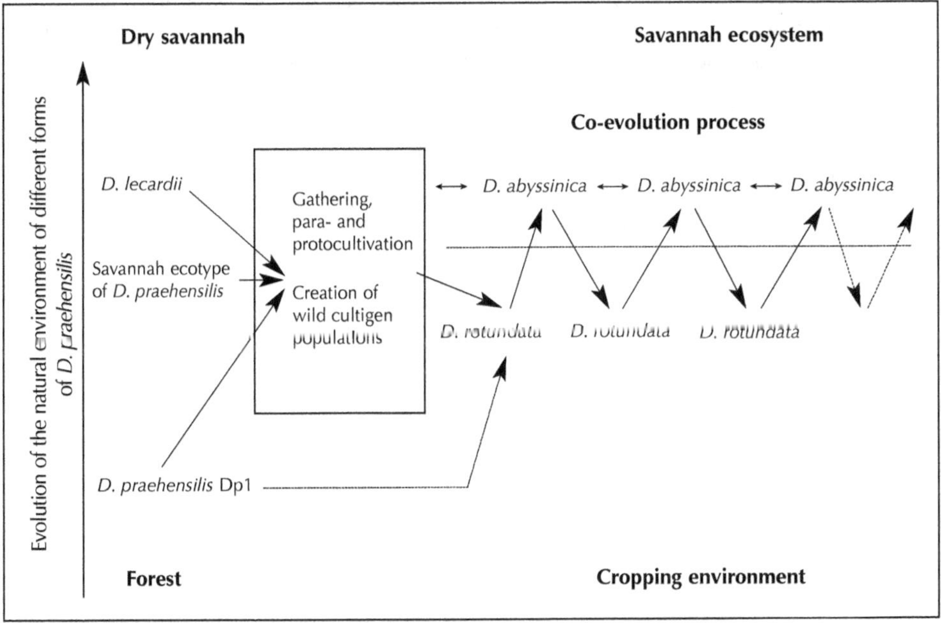

Figure 5. Hypotheses on the domestication process leading to *D. rotundata* yams.

The most common aim of such domestication is to obtain morphotypes resembling already widely cultivated cvs but which are regarded as unsatisfactory for various reasons. As often mentioned by Beninese farmers (Baco, 2000), to have a reasonable chance of success, there must be a wild form closely resembling each double-harvest *D. rotundata* cv involved in the domestication process. This would enhance the possibility of homogamy between the cultivated plant material and the wild counterpart.

The duplicate product sought by the domesticator might therefore be derived from a population of wild yams that have been highly conditioned by the coevolution process.

We do not believe that such copies can originate from cvs regrowing spontaneously in old plots. This is a very unlikely scenario in traditional agriculture, as yams are always cultivated in rotation with various crops over a period of several years. In addition, the fallows that grow after the cropping cycle are unsuitable for yams as the initial regrowth, i.e. mainly annual grasses, provides few natural stakes and is regularly subject to bush fires. Conversely, as already indicated, the fallows provide a favorable environment for natural seedlings of wild and cultivated yams once the forest regeneration stage is reached. As Mignouna and Dansi (2002) stressed, it is highly likely that the Bayere yams that they found growing spontaneously in fallows within a bimodal rainfall area were remnants of past yam crops. Furthermore, these yams originated very likely from seed carried by wind from *D. rotundata* yam crops rather than from tubers that had been left in the ground at harvest and had subsequently survived the unfavorable conditions.

The domestication of double-harvest yams is not just a matter of refining already cultivated plant material, it also occasionally generates new cvs. This is rare in northern Benin, as already discussed. There appears to be less likelihood of obtaining innovations in the geographical area where domestication has been going on the longest, perhaps because of higher genetic proximity between the cultivated yams and their wild parents.

Vine of dika yam (*D. abyssinica*).
(Photo P. Vernier)

Aerial part of a dika guea yam.
(Photo P. Vernier)

Fruits (infructescence) of a *D. praehensilis* yam.
(Photo P. Vernier)

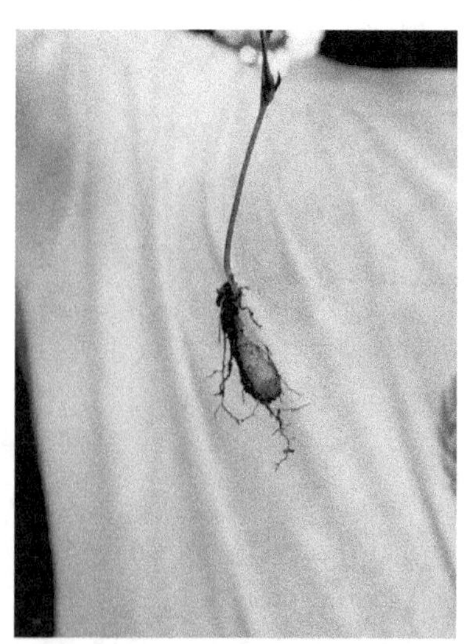

Dika plantlet grown from seed
(*D. praehensilis*).
(Photo P. Vernier)

Yam domesticating farmer with a dika guea
yam he has dug up. Sonoumon village,
northern Benin. (Photo P. Vernier)

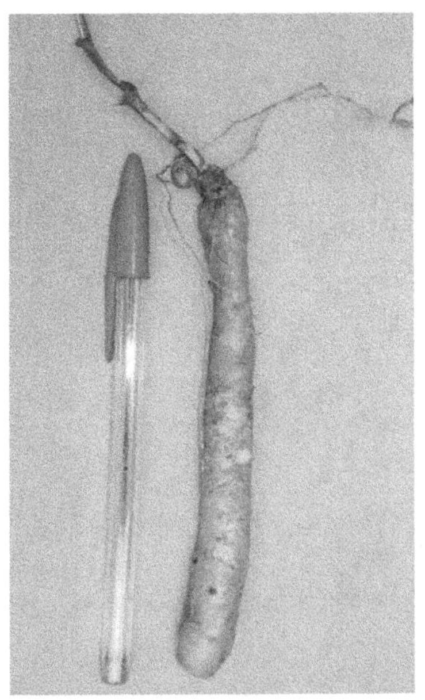

Juvenile tuber of a 'dika sinrou konsi' yam (Da1 *D. abyssinica*, cannot be domesticated).
(Photo H. Chaïr)

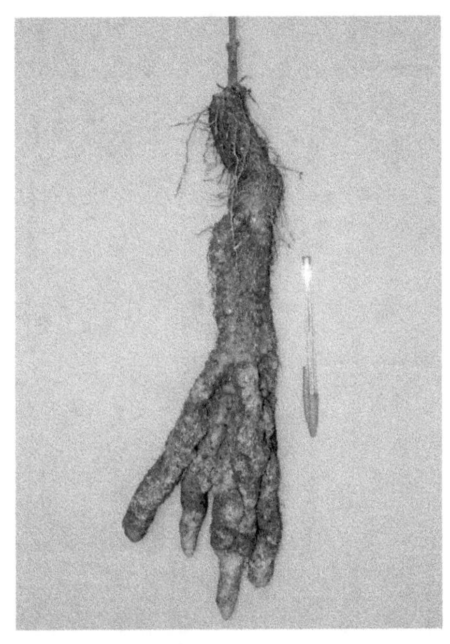

Tuber of a dika guea yam (Da2 *D. abyssinica*, can be domesticated).
(Photo H. Chaïr)

Juvenile tuber of a Dp1 *D. praehensilis* yam (cannot be domesticated).
(Photo H. Chaïr)

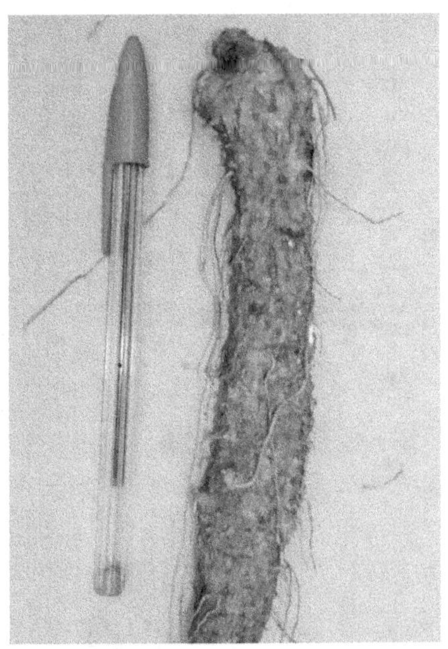

Tuber of a Dp2 *D. praehensilis* yam (can be domesticated).
(Photo H. Chaïr)

(Photo P. Vernier)

Tuber deformation after an obstacle has been placed under the seed-tuber in the mound. This kind of deformed growth indicates that this particular tuber cannot be domesticated.

(Photo P. Vernier)

(Photo P. Vernier)

(Photo P. Vernier)

Pieces of a *D. praehensilis*
tuber before and after 3 years
of domestication
(Cross River State, Nigeria).

(Photo P. Vernier)

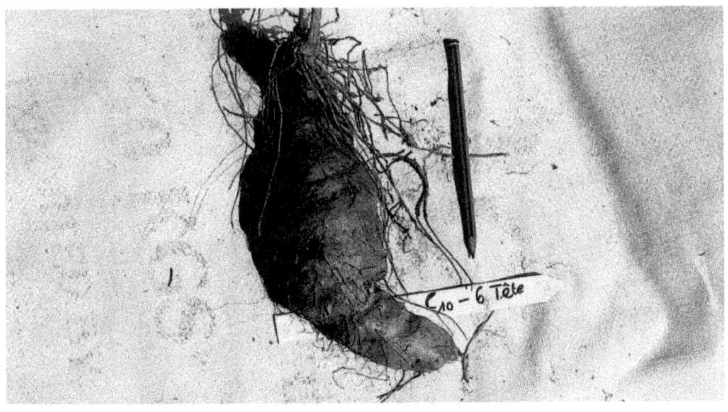

Dika (*D. abyssinica*) yam after 1 year of domestication.
Sonoumon village, northern Benin.
(Photo P. Vernier)

D. togoensis yam.
(Photo P. Vernier)

Tuberization of kokoro cultivars
(late-maturing *D. rotundata* yams
producing several small tubers).
(Photo P. Vernier)

Second-harvest tuberization of a cv
Morokoro yam (early *D. rotundata* yam).
(Photo P. Vernier)

Oufegue cultivar (hexaploid form), Cameroon.
(Photo R. Dumont)

Oufegue cultivar, reversion to the wild form, Cameroon.
(Photo R. Dumont)

Tuberization from the basal plate of a *D. burkilliana* yam
(left, R. Dumont, accompanied by Ivorian researchers), Côte d'Ivoire.
(Photo J. Zoundjihèkpon)

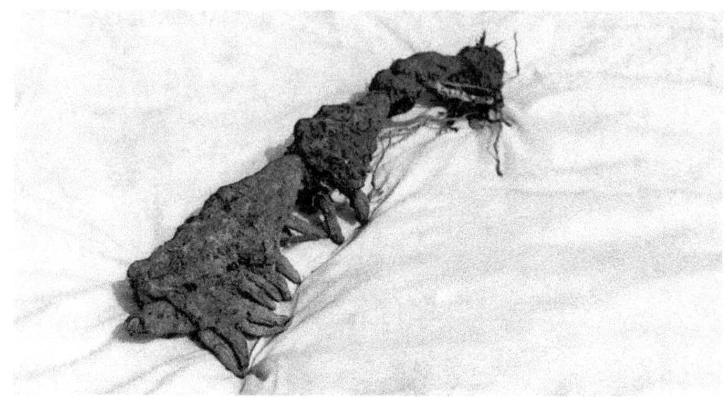

Basal plate of a *D. burkilliana* yam without tuberization.
(Photo R. Dumont)

Futu preparation in a village in Kwara State, Nigeria.
Suitability for futu-making (pounded yam) is an
important quality criterion for yam cultivars.
(Photo P. Vernier)

Cultivation techniques used to transform wild yams into double-harvest *Dioscorea rotundata* yams

General

Tuber development in wild *D. abyssinica* and *D. praehensilis* yams has to seriously compete with the biomass investment required for the aerial vegetative organs and sexual reproduction. Domestication techniques aim to reverse this situation while also improving the overall shape of the tuber and, if necessary, reducing its lengthwise growth. The adaptation of wild yams to agriculture is thus dependent on modification of their physiological functioning and corresponding morphological transformations.

Farmers' domestication work does not directly involve yam seeds. The latter are produced via natural allogamous pollination and give heterogeneous progenies, the great majority of which are aligned with wild yam morphotypes. As mentioned earlier, Camara (2001) showed that the seeds of *D. rotundata* yams differed genetically from the maternal genotype but were close to wild *D. abyssinica* yams.

Fifty years of agronomic research on the genetic improvement of yams in Africa has highlighted the difficulties encountered when using the seeds of double-harvest *D. rotundata* yams as the starting material for selection. The approach has only a small chance of success, so many progenies have to be closely examined and the result cannot be evaluated until the juvenile traits have disappeared. Recombinations between progenies then require a long waiting time, with three to five successive cropping cycles necessary to obtain a sexually functional plant from seed.

African farmers from Guinea to Cameroon bypass problems associated with using true seeds. Instead they use the tubers of old wild plants as the raw material for domestication. These are phenotypically stable and can therefore be sorted on the basis of visual and organoleptic criteria.

Domestication is usually a two-stage process. The first stage involves cropping yams collected in the wild. The morphotypes are then remodeled using specific cultivation techniques. The transformations successively obtained are not sexually transmissible but they are maintained by vegetative reproduction: a piece of tuber for cropping and a fragment of stem or even petiole (Fautret, 1985) for micropropagation.

Domestication process and effects

Domestication always begins with a choice from amongst the diversity of wild yams found in the natural flora within the vicinity of the cultivated area. The choice is based on several different criteria. Firstly, the plant must be old. It is hard to be precise, but all wild yams used for domestication seem to be derived from seeds that germinated more than 10 years earlier. This is the length of time usually required for the plants to overtop the shrub vegetation and for photosynthesis to be fully operational, thus enabling their phenotype traits to be expressed. An exception was noted in the Malinke region of Upper Guinea, in the village of Mounou, 90 km southeast of Kankan. Here domestication starts from *D. praehensilis* or *D. abyssinica* tubers that have grown from seeds 1 or 2 years earlier. The tubers are cultivated in calabashes for 2 to 3 years, then transferred to the field if they have become short, bulky and evenly shaped (Dumont, 1993).

African farmers usually look for new cvs, unusual features (large number of tubers) or morphological resemblances to the double-harvest *D. rotundata* cvs that they have already grown for a long time. This latter concern is most common in areas where domestication is a long-standing practice. In all cases, the tuber flesh must have a low bitterness and mucilage level. Wild yams are sometimes sorted on the basis of last two criteria by tasting a piece of the tuber when it is dug up. Farmers also often collect the tubers of yams recognized as 'having a sweet taste' when eaten the previous year. *D. rotundata* yams are thus, right from the start of domestication, subjected to selection with the aim of modifying the morphotype and chemotype.

Simply cultivating wild yams is obviously the most basic domestication method. However, it still represents a drastic change of ecosystem, with major repercussions on the morphological and physiological traits of the plant material. Three combined factors modify the vegetative organs. Firstly, cultivation begins with cutting up of the tuber collected in the wild. Often only the head (proximal part) is kept, but this is not a general rule. The aim is to start domestication with a small seed tuber. This represents an initial pressure that restricts the development of the aerial organs as it reduces the reserves available for growth and accelerates leaf development, with corresponding transformation of the aerial architecture. Secondly, tuber germination is delayed by several weeks, as the mound remains a relatively dry environment at the onset of the rainy season. The time available for growth of the vegetative organs is further reduced because of the yam's photosensitivity (Vandevenne and Castanié, 1988; Okezie *et al.*, 1993), with a concomitant decline in vegetative growth. Lastly, agriculture enhances the effects of the latter two factors by allowing the yams to receive the light they require without having to grow longer stems, as land clearance eliminates competition from natural shrub vegetation. The wild yams are also in much better chemical and physical fertility conditions as they are planted in large mounds of soil with a very high organic matter content. Under such conditions, the metabolic activity of the yam plants increases while the requirements of the vegetative organs are reduced. This imbalance promotes tuber development.

As already mentioned, major morphological modifications were observed in *D. praehensilis* yams subjected to environmental changes under experimental conditions (Chikwendu and Okezie, 1989). Spontaneous transformations of the same type are used by the Dourou of northern Cameroon to create double-harvest cvs from local *D. abyssinica* yams, linked earlier to *D. lecardii*. The yams are first grown in very fertile conditions, often within the village. Those reacting favorably to the change of ecosystem are propagated and then transferred to the fields. This is how cv Ngan is obtained (Seignobos, 1997). It is a yam appreciated for its earliness, although the tuber is narrow and usually weighs less than 2 kg. In southwestern Ethiopia, cvs are obtained by simply replanting wild yams of the Enantiophyllum section in family gardens (Hildebrand, 2003).

There may well be another effect underlying this architectural remodeling. When studying the development kinetics of eight *D. rotundata* cvs from Côte d'Ivoire, Zoundjihèkpon (1993) noted that most of the flower buds emerged between 15 June and 15 July, i.e. around the summer solstice in the northern hemisphere. The timing of this biological event, clearly determined by photoperiodism, is the same for *D. rotundata* yams it is as for their wild parents (Dumont, personal observation). In the cvs, however, flower primordia are formed on the branches at nodes 3 to 10, whereas they appear much further up the stem in wild plants, which need to overtop the shrub vegetation before flowering (cf. general aspects in the chapter on wild *D. praehensilis* and *D. abyssinica* yams). African annual species of the Dioscoreaceae family have a determinate growth and their vegetative development ceases at sexual maturity. In *D. rotundata* yams, shifting the plant's sexual organs to a lower architectural level is equal to stalling its development at a relatively juvenile stage. Li and Johnston (2000) provided a similar example. In the wild Cucurbitaceae *Cucurbita argyrosperma* subsp. *Sororia*, flowering occurs further down the stem after domestication. As discussed later, this physiological modification is explained by a heterochronic phenomenon.

The morphological and biological reshaping generated by cultivating wild yams does not affect the genome, as only vegetative reproduction is involved. The changes can nevertheless be sufficient to produce *D. rotundata* yams, but the grower must accept small tubers or tubers of poor cooking quality. However, this approach does not seem to be capable of producing the sophisticated double-harvest *D. rotundata* yams grown in the 'yam civilization' belt. The latter require two cultivation techniques, either separately or combined, that are designed to modify the tuber length and shape. One involves placing an obstacle, i.e. usually a piece of pottery or a flat stone, beneath the seed tuber to hamper lengthwise development of the new tubers. The other is used in the second half of the vegetative cycle. The immature tubers are removed from the mother plant before they are fully grown so as to induce the growth of new tubers of shorter length. This is the double-harvesting agricultural principle.

It is unlikely that these techniques have any symbolic relevance. They are implemented by a wide variety of ethnic groups, but they do not feature in any of the rituals that have developed around the yam. They must therefore be necessary for the transformation of wild morphotypes into double-harvest D. *rotundata* yams.

Placing an obstacle in the mound and double-harvesting are both regarded as ways of obtaining short bulky tubers from wild tubers that are long and thin. The two techniques differ in their scope of application. The use of an obstacle is associated, although not exclusively (Allomasso, 2001), with domestication in savannah areas with a monomodal rainfall regime where *D. abyssinica* yams grow. The technique is employed from Guinea to Cameroon, usually only in the first year of domestication, although it is sometimes repeated the following year or even in the following 2 years, but its role is not entirely clear. For Bariba farmers of Benin, the purpose of the obstacle is to make the tuber break its habit of growing in a lengthwise direction (Dumont and Vernier, 1997a; Baco, 2000; Okry, 2000). This is not always successful as the tubers often grow round the obstacle and then continue their normal growth. In some cases, however, lengthwise growth is halted and there is a significant increase in tuber diameter. This result, like other domestication-induced transformations, is maintained by vegetative propagation. Seignobos (1992) considered that this use of an obstacle in yam cultivation was aimed at facilitating harvesting, but a presentation published by the same author in 1997 also associated it with domestication. In simpler terms, the obstacle might be a device used to identify plants capable of modifying their natural trophic behavior. In this case, it would serve as a clonal selection operation to complement the preliminary sorting of wild yams. It would be more necessary with *D. abyssinica* than with *D. praehensilis* yams since the former, as previously mentioned, develop longer tubers.

Double-harvesting is practiced wherever *D. abyssinica* and *D. praehensilis* yams are transformed into *D. rotundata* yams. The technique is not strictly linked to the initial domestication phase. Firstly, double-harvesting may occasionally be unfeasible because the tuber is not sufficiently developed at the customary first harvest time (late July-early September in the latitude 8–10° N zone). In addition, it is possible to domesticate early-maturing *D. rotundata* yams without using double-harvesting, as shown by examples in the Malinke region of Guinea (Dumont, 1993) and northern Cameroon (Seignobos, 1997). However, the technique is always repeated annually once the yams are cropped. The agronomic role of double-harvesting is generally just mentioned even though it has two major effects. Firstly, the initially programmed biological process is interrupted, so the plant then has to rely on an emergency mechanism to ensure its survival. This stress might initiate and/or maintain deregulation in the phenotypic expression of some genes. Secondly, the new tuber (second-harvest tuber for seed use) starts developing a few weeks before the end of the rainy season and is therefore a young organ, unlike a tuber resulting from an undisturbed vegetative cycle. With sexual maturity reached at an early vegetative development stage, an additional pressure would induce 'rejuvenation' of the plant. Moreover, there is an effect of repeated vegetative propagation (tubers are a type of underground stem), which according to Hartmann *et al.* (1998) is also responsible for a progressive return to the juvenile state. There would thus be three processes that stall somatic aging of the plant.

The domestication of double-harvest *D. rotundata* yams usually takes between 3 and 5 years (Dumont and Vernier, 1997a). By the end of this period, an as yet undetermined percentage of the wild yams have acquired *D. rotundata* morphological traits. The cvs obtained are in most cases morphologically close to domestic cvs. As already indicated, exceptions are rare and tend to occur only in areas of recent domestication.

The final domestication stage is a waiting period of a few years. This allows the cultivation practices enough time to complete the morphological transformation of the product which, in parallel, is assessed in terms of different agronomic and culinary aspects. At the end of this stage, if the domestication result is a recognizable copy, its tubers are mixed with those of the reference cv and it is referred to by the same name. Similar behavior was observed by Hildebrand (2003) in southwestern Ethiopia: "Farmers [...] relate certain wild-growing and domestic varieties to one another. If an adoptive transplant proves to resemble a domestic variety, they may give it the name of the domestic variety a few years after transplantation."

The above practice accounts for the polyclonal nature of cvs reported in several studies (Hamon, 1987; Dansi *et al.*, 2000b; Mignouna and Dansi, 2002). This genetic diversity can be sorted, as carried out for cv Krengle of Côte d'Ivoire between 1987 and 1990 (Tokpa and Dumont, 1995). Using 500 'head of line' tubers collected in markets, an equivalent number of clonal lines were created by vegetative propagation. The selected sample consisted of 72 specimens with short tubers and no epidermal roots. The sample probably still contained some genetic variability, as the yields of the different lines ranged from 8.5 to 15 t/ha. It is not known whether this reflected a difference in intrinsic productivity or a variability in resistance to yield-reducing factors.

Certain domesticated yams seem to be approximate copies of already existing cvs. They occasionally produce variant cvs, several examples of which were encountered in northern Benin (Dumont and Vernier, 1997a; Dansi *et al.*, 1999). In these cases, the name of the cv is followed by an identifying suffix. For instance, cv Boni Oure is divided into about 10 types, distinguished by differences in tuber shape: Boni Oure bagarou ('big tuber'), Boni Oure ketekoba ('tuber curved like a cow horn'), Boni Oure woloukaba ('tuber like the root of the shea tree (*Vitellaria paradoxa* Gaertn)'), etc. In other cases, the variant cv is given a new name. Douroubayesirou and Kpakara cvs are thus variants of the now endangered Maraworoukorou and Danwari cvs. Bariba farmers have an ambivalent attitude towards variant cvs, but there is still not enough information available to explain this. They also give completely new names to cvs representing real innovations, e.g. Toko nou woura and Spatake, mean 'if the stranger accepts you' and 'the yam that you do not sell', respectively (Bako, 2000). A similar attitude was noted among Nago and Fon domesticating farmers in southern Benin (Mignouna and Dansi, 2002).

Some domesticated yams cannot be incorporated with already existing cvs or do not give rise to variant or new cvs of obvious agronomic interest. They are kept on the farm (often until the death of the farmer who has created them) but are not propagated. Such yams appear to be failures, but this needs to be qualified. Some kokoro yams, which are now extremely important for the sedentarization of African agriculture and a food product that is extremely well adapted to the urban market, are derived from once neglected cvs. Domesticated yams previously of no interest can therefore play a key role in development if their production is reoriented in response to a combination of socioeconomic changes and technical constraints, particularly more intensive use of available land.

Thoughts on domestication leading to double-harvest *Dioscorea rotundata* yams

Domestication appears to have two main improvement effects at the cv level. Firstly, greater clonal diversity, sometimes with readjustment of its homeostatic equilibrium. Secondly, a possible inheritance of beneficial traits, as clones obtained from domesticated yams are retained only after selection through several cropping cycles.

The involvement of sexual reproduction in the wild yams likely means that morphotypes familiar to the farmers can be obtained, but the genotypes will differ in terms of at least some genes. From this standpoint, double-harvest *D. rotundata* yams can be seen as the result of a series of subtle readjustments, each leading to genotypic innovations. This seems to be a long-term process. For instance, cv Soussou from the Bariba region of Benin studied by Chevalier in 1920 is still being improved by domestication (Dumont and Vernier, 1997a; Baco, 2000).

D. rotundata yams thus evolve genetically, but the initial morphotypes remain unchanged or undergo modifications that are barely perceptible on a human generation scale. Each *D. rotundata* cv consists of several clones of different genetic ages, which could bear different biological, agronomic or culinary traits. It is a pool of modest evolutionary variability that can be useful as a source of adaptations when required after a change of environmental conditions or the emergence of new agricultural constraints.

This evolutionary capacity was already suspected by Hamon (1987) and Adoukounou (2001). Its functioning and potential benefits obviously cease when the plant material is no longer domesticated due to a loss of sexual reproduction capacity. This suggests that non-floriferous cvs have little genetic diversity. This genetic impoverishment seems nevertheless to be a very slow process. In Benin, the Agogo, Boni Oure, Kokouma, Morokorou and Soagona cvs are still highly polyclonal (Dansi *et al.*, 2000b) and the same is true of cv Bakokae of Cameroon (Mignouna *et al.*, 2002b).

Domestication is organized around sexually functional cvs, so it is focused on plant material with a secondary role in present-day agriculture. A study of a Bariba village in northern Benin (Dumont, 1997) revealed that floriferous cvs occupied less than 20% of the area cropped with double-harvest yam cvs. Similar percentages have been reported in other studies conducted in the same ethnic region (Dumont and Vernier, 1997a; Baco, 2000; Vernier *et al.*, 2003).

In the part of the Bariba region of Benin studied by Vernier and Dossou (2000), low-fruiting Morokorou and Kokouma cvs occupied over 50% of the area under double-harvest yams, which in turn accounted for a third of the total area under yams. In northern Cameroon, cv Bakokae (Seignobos, 1997), which seldom fruits, is in an even more dominant position. The situation is very different in Côte d'Ivoire, where yam production is shared regionally between several double-harvest cvs, all of which are floriferous (Hamon, 1987). This situation might be linked to the fact that *D. rotundata* yams have probably existed there for a shorter length of time.

The above observations suggest that sexually functional double-harvest yams are the ones that have most recently germinated from seed. There are two arguments in support of this idea.

Firstly, farmers currently only bring together sexually active cvs via domestication, thus, by extension, making a link with the sexual features of wild yams (Dumont and Vernier, 1997a). Moreover, earlier studies on *D. rotundata* (Sadik and Okereke, 1975; Zoundjihèkpon, 1993) indicated that passage through the germ line enhances the sexual potential of yams. Domestication seems to be focused on ever-evolving plant material. The situation noted today with respect to the concerned cvs would be a new episode in a series of attempts, often made in the very distant past, to gradually improve still sexually active and perfectible cvs. This interpretation is in agreement with the prime concern of farmers not to shift from their traditional morphotypes.

Secondly, there would be a cause-and-effect relationship between the introduction of sterility and the improvement in agronomic value. A major energy investment is required for sexual reproduction at the expense of the tuber. This is particularly true in female plants. As a first approximation, their infructescence mobilizes 0.5 kg of dry matter in *D. abyssinica* and even more in *D. praehensilis* (Dumont, personal observation). It has already been pointed out that impairment and interruption of flowering are basic trends in double-harvest cvs. It is not known whether these phenomena are the normal prolongation of domestication (possible effects of repeated vegetative reproduction) or if they originate from a more erratic event (sexuality-inhibiting mutations) that is beneficial for the farmer. Nevertheless, the cvs concerned are disconnected from the coevolution process and the farmer cannot improve them further.

There would thus be two categories of double-harvest *D. rotundata* yams. Firstly, cvs that are no longer sexually active. These generate what is now the most productive plant material, but they are threatened with decline in the long term because of their incapacity for adaptive genetic diversification other than via potential favorable mutations. Secondly, cvs that are in the domestication process, which will likely supplant yam cvs in the first category in the distant future. With such plant material, farmers benefit from past achievements while at the same time preparing for the future.

On the basis of previous considerations, the conditions favorable for domestication would be optimized by coevolution dynamics that increase the genetic similarity of wild *D. abyssinica* and *D. praehensilis* yams relative to *D. rotundata* cvs. Sorting among the thus modified wild populations would provide plant material capable of transformation. Overall, this is essentially a genetic process in which sexuality plays a important role. In addition, specific domestication techniques are required to achieve a *D. rotundata*-like product. These are directed stimuli applied under specific agricultural conditions. They produce an effective result only with genotypes that can modify their phenotype expression in the direction desired by the domesticating farmer. The origins of *D. rotundata* yams are thus limited to *D. praehensilis* and *D. abyssinica* yams capable of responding to domestication techniques. Here again the genetic aspect is important.

Transformations induced by domestication are not genetically determined and the mechanisms that might explain them are still very hypothetical. Ontogenetic or biological variations not directly related to the genome appear to play a major role. It has already been stressed that several techniques are used to induce and maintain an artificial state of relative juvenility in *D. rotundata* yams, and that these techniques are repeated. This artificial juvenility widens the gap between the plant's chronological age and its physiological age. The genotype would be disconnected from the phenotype, with a germen to soma shift in the programming of ontogenetic development. This is in agreement with the highly innovative idea of Hallé (1999): "The hypothesis that the shape of the plant is established without involving the genes seems to be currently gaining ground among biologists. [...] It would be as if the plant shape carried a sufficient amount of information to replace the genome." This idea is similar to the memory concept mentioned later and, in another sense, it would explain why domestication-induced traits are not sexually transmissible but are maintained by vegetative reproduction. However, none of this accounts for the morphological or other transformations brought about by domestication.

Porthig and Scott (1998) believed that the characteristics of the organs produced during the different plant development phases were regulated by two independent programs, one modulating morphogenetic and cellular differentiation and the other controlling the duration of organ formation. A change in the functioning of either of these programs would correspond to a heterochronic mechanism generating morphological and physiological diversity.

Heterochronic phenomena are best known at the macro-evolutionary level. They have been widely studied in the animal kingdom but have received far less attention in the realm of plants. However, Li and Johnston (2000) published an excellent synthesis of current knowledge, presenting the various forms of heterochrony and their respective contributions to the evolutionary processes of plants. Their work provides no information specifically relating to root and tuber plants, but two relevant ideas should be noted. Firstly, that heterochrony can occur at the intraspecific level, generating variations in the shape and size of different organs of the plant or modifying its physiological functioning. Secondly, that the appearance of heterochronies can be triggered by external factors that disturb the plant's normal functioning.

The above ideas apply very well to the domestication of *D. rotundata* yams, a large part of which involves exerting powerful and apparently continuous pressure on the wild forms. The juvenile state of the plant would be obtained and maintained by a particular heterochronic mechanism, called 'progenesis' by Li and Johnston (2000) and 'hypomorphosis' by Chaline (1999). According to the latter, the disturbance involved here would be of epigenetic origin. Concerning yams, the disturbance could be initiated by the drastic biotope change caused by domestication. Unlike the natural soil environment, which is protected from sunlight by the vegetation, the mound accumulates a large amount of heat, creating an unusual or even traumatizing environment for the seed tuber. According to Chaline (1999) "Heterochronies are often associated with fluctuations in climatic parameters via heat-sensitive genes, thus triggering the production of hormonal mediators".

A number of cultivation practices appear to combine agronomic advantage and domestication effect. Cutting up large tubers or using second-harvest tubers as seed tubers is a health protection measure, as tubers from plants weakened by disease (including viral infections) cannot be used for seed purposes and they are even excluded from reproduction. Another example is the dual role played by the mound. Planting yams in mounds protects them (sometimes inadequately) against hydromorphic phenomena and greatly facilitates harvesting. Most importantly, harvesting can be postponed until the middle of the dry season, with a corresponding reduction in the duration of open-air storage and its associated losses. At the same time, the mound is an essential domestication tool, enabling progressive improvement in the size and overall shape of the tuber with gradual elimination of the epidermal roots.

The domestication products seem to be potentially unstable at various levels. This can be illustrated by three examples: occasional inversion of sexuality, occasional sterility in usually fertile cvs and the temporary appearance of abnormal leaf or stem traits. There are probably also risks of morphological deviations, which would explain why the tip (distal part) of the tuber is removed when seed tubers are prepared for planting. This practice is frequently observed in traditional agriculture (in Benin and Côte d'Ivoire) but the underlying reasons are unclear in the minds of the farmers. It is thought to be precautionary measure against the development of an atavistic tendency that would remodel cultivated yams according to the wild morphotypes. There has been a spectacular example of this phenomenon concerning Cameroonian yams introduced into Côte d'Ivoire (Dumont *et al.*, 1994). The reversion of West African *D. rotundata* yams to their wild forms is far less conspicuous as the two types are morphologically (and probably genetically) much closer. One such reversion involves cv Kouroutokokoragourouko (literally 'old woman who never ages') of the Bariba region of Benin, which has now virtually disappeared. If removal of the tuber tip actually controls the reversion mechanism, then the operation could disable a memory established in this part of the organ. Zoundjihèkpon (1993) had already suspected the existence of a 'somatic memory' implicitly located in the yam tuber. This brings us back to the hypothesis of Hallé (1999), which suggests that some traits of the plant might not be directly controlled by the genome.

Domestication (in the broad sense) could also be regarded as a process that hampers yam cv degeneration. Passage through the germ line is usually regarded as an effective process for eliminating viruses accumulated by vegetatively propagated plants. Allano and Clamens (2000) proposed that sexual reproduction might be a DNA repair mechanism in species that periodically use this form of reproduction: "With asexual reproduction, mutations accumulate in the population and these are almost always unfavorable [...]. Sexual reproduction eliminates them by meiosis. [...]. Alternation between asexual and sexual reproduction could therefore be considered as a kind of periodic fixing of the genome."

Arguments can be put forward against the previous hypothesis that several no longer sexually active double-harvest *D. rotundata* yams can be successfully cultivated and at times even have a dominant place in African agriculture (e.g. cv Morokorou of northern Benin and cv Bakokae of Cameroon mentioned earlier). The decline or loss of sexual function would not jeopardize the existence of the cultivated yams and their sporadically observed degeneration could be explained by other causes (cf. section Significance and importance of domestication for farmers). These possible negative pressures seem to have an extremely slow impact in traditional agriculture conditions.

The principal benefit of domestication could be summarized on the basis of two major effects. One would be the periodic contribution of clonal innovations in sexually functional *D. rotundata* yams, resulting in genetic enrichment of already existing cvs, genesis of variant cvs and, more rarely, creation of genuinely new cvs. The other would be the maintenance of all plant material under cultivation. Considered from this standpoint, traditional cultivation practices would be a necessary continuation of techniques implemented in the primary domestication phase.

In return, domestication would link the plant material to a particular biotic environment and it can be assumed that degradation of this environment would endanger the beneficial transformations obtained. *D. dumetorum* and *D. bulbifera* yams thus reacquire their toxic characteristics when no longer cultivated (Dumont, personal observations in Cameroon and Guinea). According to Beninese farmers, *D. rotundata* yams revert to their wild forms when reestablished in the savannah or forest environment (Dumont and Vernier, 1997a; Allomasso, 2001). The above ideas highlight the need for research into the potential impact of major changes in traditional cultivation techniques on the morphological stability of domesticated yams.

More fundamentally, it is still not possible to explain the biological mechanisms involved in domestication, the stability of the transformed traits under vegetative propagation and the potential fragility of this situation. This explanatory research is laborious and risky. Firstly, the available information is fragmentary and has often not been adequately verified and, secondly, the problem appears to lie beyond the scope of Mendelian genetics.

Although yam domestication techniques do not result in any genomic changes, the genome's phenotypical and physiological expression is profoundly modified. Zoundjihèkpon (1993) already suggested the existence of this phenomenon to explain interannual variations in phenology observed in a number of *D. rotundata* yams from Côte d'Ivoire. Hence, morphological transformations induced by domestication might be determined by regulator genes rather than by structural genes.

The domestication techniques used, and even the standard cultivation practices, upset the development dynamics of wild yams brought into cultivation. According to Gould and Hall (1999), this sets up an epigenetic process (i.e. one not entirely controlled by the genetic program) that is very effective in inducing morphological variations. Atlan (1999) indicated that the phenomena in question might result from variations affecting the DNA methylation status or the degree of acetylation of histones, i.e. proteins associated with DNA in the chromosomes. He also presented the 'genome fingerprint' as a further example of genetic inheritability. Although they have the same structure, the expression or non-expression of certain genes depends on whether they come from the father or the mother and this differential status is conserved after germ cell division and subsequent gamete fusion. It would be interesting to test Atlan's latter hypothesis against the case of the wild parents of *D. rotundata* species, as it has earlier been seen that the two sexes do not have the same probability of passing on their genes and that domestication can change this situation.

The epigenetic phenomena might explain why certain transformations induced by domestication techniques are conserved by vegetative reproduction. Moreover, this idea is part of a much more general one. Most evolutionists these days agree that the (fine) adaptation of species occurred as a result of changes in the regulator genes (Gouyon *et al.*, 1997).

Conclusion

This is the first time that existing knowledge concerning the domestication of *D. rotundata* yams has been presented and examined synthetically. We have endeavored to formulate a series of hypotheses that could be used when comparing different theories, identifying research topics and discussing results. It represents a paradigm from which ideas can be drawn that will shape the direction of future investigations.

As we have seen, ecotype variability transforms *D. praehensilis* yams into *D. abyssinica* yams and either of these can be domesticated into *D. rotundata* yams, whose traits are then transferred back to the wild parents by gametic and zygotic migration. The three yams in question should therefore be regarded as a single species. According to international rules of nomenclature, this species would be *D. rotundata*, defined in 1813, as compared to 1849 for *D. praehensilis* and 1850 for *D. abyssinica*. This nomenclatural classification is the result of historical chance and reverses the order of the evolutionary system presented in our study.

In general, the botanical definition of the plant material used in and produced by domestication is clearly weak. It is therefore essential to clarify, or even redefine, the taxonomy of the African Dioscoreaceae of the Enantiophyllum section, while focusing especially on its dynamic dimension. To work effectively in this direction, studies must be undertaken at the population scale and in a broad range of ecosystems, so as to take the diversity of each of the yams currently regarded as distinct species into account. It would also be essential to widen the botanical scope by conducting a parallel study of the Asterotricha section, also endemic to Africa, as several arguments indicate that the latter is phyletically close to the African Enantiophyllum section. Information that could enhance our overall understanding of the present diversity of African members of the Enantiophyllum section could probably be found at the intersection of these two sections. Lastly, to establish the main lines of this diversity, it might be advantageous to first identify the maternal lines through mitochondrial DNA analysis. This initial rough assessment of the problem could then be further refined using other techniques. Note that *D. rotundata* cvs are often polyclonal, so the range of this variability should be determined before attempting to structure it and identify its connections with wild yams. The development of a tool to construct a genealogical hierarchy of yam cvs and their clonal diversity would also be a major scientific advance, with obvious practical applications. In short, the investigation methods need to be enhanced in order to boost the productivity of yam genetics investigations.

We suspect that the *D. abyssinica* yams of West Africa and, on a far wider geographical scale, the Dp2 form of *D. praehensilis*, are cultigen populations (probably in many cases genetically compatible) that contain genetic combinations which have sometimes developed over centuries of human manipulation and are responsible for their evolution. These botanical groupings should be viewed as an African or even a world heritage. Their disappearance would be a major technical, cultural and scientific loss, as they represent one of the rare processes of plant domestication still observable today. It is therefore crucial to determine ways to protect them. An essential prerequisite in this regard is the implementation of conservation measures capable of maintaining long-term fallows, the genetic diversity of *D. rotundata* yams and ancestral agricultural techniques. Similarly, the many sacred, and often endangered, forests still found in West Africa also need to be preserved. These are sanctuaries of biodiversity within which the relicts of earlier populations of wild or even cultivated yams can continue to survive.

There are regional variants of the domestication technique, one specific to the area of *D. rotundata* yam civilization and two (in Guinea and Cameroon) on the fringes of this area. Each of the latter two cases is based on a single observation and requires confirmation. In addition, the domestication techniques need to be listed in greater detail, as some practices have probably been overlooked in our study. It would also be useful to experimentally test the currently known techniques in order to assess the potential and advantages of including them in a modern genetic improvement process.

The genetic improvement of *D. rotundata* yams by sexual reproduction is a difficult challenge to meet. If these yams are as described in the previous chapters, there are few opportunities for traditional selection via intercultivar recombinations. Flowers of single-harvest cvs are probably sterile. As far as double-harvest yams are concerned, the agronomically productive cvs cannot be genetically manipulated since they are no longer sexually functional. In addition, each of the few regularly floriferous cvs would correspond to a selection process that is still under way within the framework of traditional domestication, and they would actually be imperfect products. These would consist of clones of different genetic ages and probably varying agronomic and cooking quality. In the light of all these difficulties, it is understandable that African farmers prefer to rely on wild yams. This means that the coevolution process mentioned several times earlier must function effectively. This hypothetical phenomenon would be central to the domestication issue, and its relevance will likely soon be confirmed if research is continued on this subject.

A dimension of genetics beyond Mendelian theory probably has to be taken into account to understand the different factors involved in domesticating *D. rotundata* yams. The capacity to respond to domestication techniques would still depend on the genome. However, traits induced by domestication and subsequently selected would be controlled by the soma, which is restored to and maintained in a juvenile state. This double hypothesis has still not been taken into account when designing programs for the genetic improvement of African yams. In the light of all the above considerations, we propose a number of lines of research that could be investigated with the aim of developing a technique to improve *D. rotundata* yams.

In general terms, it appears that any operation aimed at genetically transforming *D. rotundata* yams should be preceded by clonal sorting of their parents. If this is not done, there is a risk of using obsolete or inappropriate genotypes.

Selection via sexual reproduction is an interesting technique despite the few results obtained to date. However, two conditions are needed to make it more productive. Firstly, a wider range of parents is essential. This requires either the recovery of sexuality by no longer sexually functional cvs or the use of sexually functional cvs identified through a survey of the area in which *D. rotundata* yams are grown. Secondly, selection via sexual reproduction should be combined with traditional domestication techniques, especially double-harvesting and the use of an obstacle in mound cultivation, both of which have proved their worth in the selective sorting and transformation of wild yam material.

The large-scale introduction of sexually functional foreign cvs into local agriculture should modify the genetic equilibrium of *D. abyssinica-D. praehensilis* yams in the surrounding environment and in return enable clonal innovations to be obtained by traditional domestication. The method is easy to implement but its results are unpredictable and take a long time to achieve.

If the agriculturally useful traits are controlled by the soma, they can hopefully be transferred horizontally by stem grafting, if generalizable, combined with the use of bacterial or possibly viral inoculants. This would provide an alternative to protoplasm fusion, which has been used experimentally over the last few years to obtain somatic hybridization. In principle, both techniques would allow intraspecific combinations that are not presently possible because many cvs are either no longer floriferous or sterile. Interspecific combinations between *D. rotundata*, *D. alata*, *D. cayenensis* and other cultivated or wild species should also be possible.

Finally, our study of domestication incidentally includes considerable information on various aspects of *D. rotundata* yams that come under different research domains:

Yams are clearly a major asset for African agriculture. They are proving to be plant material of remarkable genetic richness and tremendous biological plasticity. As little is known about this botanical capital, it would be essential to conduct an exhaustive survey of the diversity of West African *D. rotundata* yams. At the same time, the know-how of local farmers relating to the use of these genetic resources should be document further, particularly the technical innovations developed to adapt them to modified cropping systems and commercial constraints. Many traditional skills in these areas would probably be useful in agronomic research and rural development.

The medicinal properties of African yams have been mentioned several times in this book. In-depth scientific research into this relatively unexplored subject is being carried out in Latin America and to an even greater extent in Asia, particularly in China and Japan, and there have already been practical spinoffs. Africa undoubtedly possesses similar resources and a better understanding of African yams would foster their development.

Like *D. cayenensis* yams, *D. rotundata* yams are deeply rooted in the cultural universe of rural African people. Information on this sociological environment needs to be rapidly collected, as very ancient traditional knowledge is threatened with disappearance if its present erosion continues. In particular, cultivated yams are excellent markers of human migration, so their history should be a useful source of information for ethnobotanists and more generally for specialists in human sciences with an interest in sub-Sahelian Africa.

Glossary of technical terms

Most of the definitions are translated from the *Dictionnaire de Génétique*, ed. J.-C. Sournia (1991).

Allelic mutation (formerly genetic mutation): Mutation affecting a single or a limited number of bases at a given locus. Creates a mutant allele.

Biodiversity: Contraction of 'biological diversity' proposed by the National Research Council (NRC) after its first forum on the subject in Washington in 1986. Biodiversity is the overall diversity of all living organisms, considered on three levels: ecosystems, the species within ecosystems and genes present in each species.

Biotope: Minimum area in which a species encounters the environmental conditions that it requires. By extension, all physical and chemical factors characterizing an ecosystem.

Chemotype: Cultivar distinguished by a chemical trait. The polyphenol concentration is the main chemotypical criterion separating *D. rotundata* yams.

Chromosome mutation: Mutation affecting either the structure of one or more chromosomes or the number of chromosomes or both.

Climax: Plant community in equilibrium with the general climate. A succession of communities leading to the establishment of the climax can be a progressive, e.g. fallow regrowth, or regressive, e.g. evolution of forest to savannah, series (as defined by Schnell, 1971).

Clone: Set of genetically identical individuals derived from a sexed embryo by asexual reproduction.

Codon: Triplet of nucleotides coding for a single amino acid.

Cultigen: Term used as a noun or adjective in scientific literature. We have been unable to discover its origin. It was used by Burkill in 1939 and Miège in 1952. It was then adopted by different authors of publications on yams. In the broad sense, it is applied to wild yams that are phenotypically modified as a result of cultivation. We extend the idea of cultigen to populations of wild yams whose natural equilibrium is broken by human intervention, either by giving preference to useful genotypes or by developing an agricultural system that promotes gene flows capable of 'reformatting' the genetic diversity of the populations concerned.

Cultivar: Contraction of cultivated variety. Entity formed from various genotypes of the same species resulting from mass selection within the framework of traditional agriculture. This definition implies that cultivars are polyclonal, as is virtually always the case with *D. rotundata.*

Dioecism: Condition of a plant species whose individuals are unisexual.

Dried chip: Yams are processed into chips (or 'cossettes') to prolong their storage life. Whole tubers or tuber pieces are peeled, if necessary sliced, parboiled in water (for 30 min at around 70°C) and sun-dried. About 25 to 35 kg of chips with a 12% moisture content are obtained per 100 kg of fresh tubers. The varieties most commonly used for this purpose are single-harvest kokoro-type *D. rotundata* yams. Once dried, the chips are very hard and have a long storage life (1 year or more) provided that they are protected from weevils. They must be milled into flour before preparing the dough (amala) or couscous (wassa-wassa).

Ecosystem: Dynamic complex consisting of communities of plants, animals and micro-organisms and their physical and chemical environment that interact to form a functional entity.

Ecotone: Transitional area between two different ecosystems, as is the case with the forest/savannah interface.

Ecotype: Set of individuals of the same species that occupy the same biotope and whose adaptation is the result of selection by this particular environment.

Epigenetic: Refers to changes in the transcriptional status of genes. These are heritable through cell division but do not involve changes in the DNA sequence. Such traits are usually reversible, depending on the cell type, and are the result of what are called epigenetic modifications. One of the best-known of these modifications is DNA methylation (addition of a methyl group), often reflected in the inhibition of gene expression (i.e. gene silencing).

Gene: Nucleotide sequence forming a unit of genetic information and determining the expression of a trait either directly, though a structural gene, or indirectly, though a regulator gene. This definition needs to be qualified. Several genes can contribute to the expression of a trait, and the same gene can contribute to the expression of several traits. More generally, the definition of a gene is still a subject of controversy.

Genetic variability: Estimate of the genetic diversity within a population or a sample using quantitative genetics methods involving estimation of means, variances, covariance, etc.

Genome: Complete set of genes present in a virus, organelle or single cell organism or in the cells of a multicellular organism which program and control its structure, functioning and development.

Genotype: Set of genes of an individual, whether expressed or not, revealed by genetic or molecular analysis.

Geophyte: Plant with underground perennating organs (rhizomes, bulbs, tubers).

Haplotype: Genetic profile of chloroplast DNA. Individuals with a common haplotype belong to the same species.

Heterochrony: Change in the timing and rate of development, modifying the progeny of an ancestral species. Such changes are the result of genetic processes controlling development.

Heterodimer: Molecule resulting from the combination of two different molecules.

Homeostatic equilibrium: Genetic standardization of a population. Under constant environmental conditions, all genes are stabilized at equilibrium frequencies reflecting their respective contributions to the overall adaptive value. The homeostatic equilibrium of populations of wild yams giving rise to *D. rotundata* yams can be broken either by a sustained climatic change or by anthropogenic disturbances such as a reduction in the population or a modification of gene flows induced by agriculture, as a result of a particular constraint (low fertility, commercial requirements) that reduces the diversity of the cultivated plant material.

Homodimer: Molecule resulting from the combination of two similar molecules.

Homogamy: Mode of sexual reproduction in which the individuals in a population do not mate randomly, but rather according to phenotype resemblances.

Hypomorphosis: Advancement of sexual maturity, arresting the growth of an individual and interrupting its development at a relatively juvenile stage.

Microevolution: Set of gradual changes occurring over time within a species and resulting in better adaptation of the individual to its environment.

Monoecism: Condition of a plant species whose individuals are bisexual and have unisexual flowers.

Morphotype or Morphovar: Variety distinguished by a specific morphological characteristic.

Mutation: Naturally occurring or induced modification of the genome, or of a cell or organism. Usually hereditary.

Panmixia: Mode of sexual reproduction in which each individual in a population has an equal probability of mating with any individual of the opposite sex belonging to the same population.

Paradigm: Major theoretical shift resulting from an in-depth change of vision by a researcher or a community of researchers who view the facts in a fundamentally different way from their predecessors (definition of Coppens and Picq, 2001).

Parthenocarpy: Process during which a fruit develops that is either seedless or contains seeds with no embryos.

Pauciflorous: A plant with few flowers as a result of inflorescence atrophy or a decline in budding, both of which occur in *D. rotundata* yams.

Perennial basal plate: Lignified mass surmounting the tuber of *D. cayenensis* yams and of most wild species of the Enantiophyllum section with pluriannual or perennial vegetative organs. It contains the nodal complex, the site of germination. In *D. praehensilis*, *D. abyssinica* and *D. rotundata* yams, the basal plate is reduced to a pre-tuber (sometimes called a corm) or is absent, in which case the nodal complex is located in the tuber head.

Phenotype: Set of observable traits of an individual or group resulting from the expression of the genotype in a given environment.

Premature tuber formation: Form of morphogenesis affecting yam tubers when stored in the dark at high ambient humidity. There is a proliferation of parenchymal tissue adjacent to one of the dormant buds remote from the apical buds. A daughter tuber develops from the mother tuber, thus short-circuiting the vegetative phase. The phenomenon is repetitive and has long been used by Bariba farmers of Benin for long-term storage of cv Boni Oure (2 years or longer, they claim), obviously with very substantial weight loss.

Progenesis: See hypomorphosis.

Regulator gene: Gene whose product is a regulator protein controlling the expression of structural genes.

Sacred forest: Area of natural vegetation maintained for ritual practices, often dating back to the ancient past.

Species: Set of individuals whose reciprocal fertilization produces fertile seeds giving rise to progeny that resemble the parents and are capable of interbreeding without losing their morphological stability. This definition is hardly applicable to *D. rotundata*. Its morphological stability is not maintained by sexual reproduction. In addition, single-harvest *D. rotundata* yams are probably sterile and numerous *D. rotundata* cvs have more or less lost their sexual reproduction capacity.

Structural gene: Gene whose product is RNA, tRNA, rRNA, a structural protein or an enzyme.

Tropophyte: A plant adapted to alternating climatic conditions of rainfall and drought, as is generally the case with savannah vegetation.

Variant: Individual with different traits from those of most of the population to which it belongs. With *D. rotundata* yams, the variant sometimes has a sufficient advantage to supplant the original *D. rotundata* population.

Variety: In taxonomy, a subdivision of the species. It is considered to differ from the type species in only a small number of clearly defined exclusive traits. With *D. rotundata*, the variety (like the cultivar) is not a natural sexually reproducing population. It is usually a mixture of clones corresponding to the same phenotype and reproduced by vegetative propagation.

Vegeculture: Cropping system based on root and tuber plants reproduced by vegetative propagation, with little tilling. Such systems are usually associated with perennial produce in complex staged arrangements. The archetype is the Melanesian or Creole garden.

References

ABRAHAM K. A., GOPINATHAN-NAIR P., 1991. Polyploidy and sterility in relation to sex in *Dioscorea alata* L. *(Dioscoreaceae)*. Genetica 83 (2): 93-97.

ADOUKONOU H., 2001. Gestion paysanne de la diversité génétique du complexe *Dioscorea cayenensis - rotundata* au centre du Bénin. DEA graduate-level thesis, Université de Lomé, Togo, 56 p + appendices.

AKISSOÉ N., HOUNHOUIGAN J., MESTRES C., NAGO M., 2003. How blanching and drying affect the colour and functional characteristics of yam *(Dioscorea cayenensis-rotundata)* flour. Food Chemistry 82 (2): 257-264.

AKORODA M. O., CHHEDA H. R., 1983. Agro-botanical and species relationships of Guinea yams. Tropical Agriculture 60 (4): 242-248.

AKORODA M. O., 1985. Pollinisation management for controlled hybridation of white yam. Scienta Horticulturae 25: 201-209.

ALLANO L., CLAMENS A., 2000. L'évolution. Des faits aux mécanismes. Collection Sciences de la Vie et de la Terre. Editions Ellipses, Paris, France, 160 p.

ALLOMASSO T., 2001. Conservation des ressources génétiques forestières du département de l'Atlantique: stratégies de conservation de l'igname sauvage *Dioscorea praehensilis* Benth dans les forêts sacrées et étude de sa domestication. DESS graduate-level thesis, Université nationale du Bénin, Faculté des sciences agronomiques, Cotonou, Benin, 88 p.

ARNOUX M., BLAYAC H., CLAPARÈDE L., 1996. Influence des facteurs du milieu sur l'expression de la sexualité du chanvre monoïque *(Cannabis sativa* L.). Action de la nutrition azotée. Annales de l'amélioration des plantes 16 (2): 123-134.

ASSOGBA C. R., 1993. Rituels de l'igname et altérité en Afrique de l'Ouest. Pour une théorie de la contracculturation. PhD thesis, Faculté des lettres, arts et sciences humaines, Université nationale de Côte d'Ivoire, Abidjan, Côte d'Ivoire, 448 p.

ATLAN A., GOUYON P. H., 1994. Les conflits intra-génomiques. Médecine Science 10 (3): I-IX.

ATLAN H., 1999. La fin du « tout génétique ». Collection Sciences en questions, INRA Editions, Versailles, France, 87 p.

AYENSU E. S., 1972. Dioscoreales. *In* C. R. METCALFE Anatomy of the Monocotyledons, vol. 6. Clarendon Press, Oxford, UK.

AYENSU E. S., COURSEY D. G., 1973. Guinea yams. The botany, ethnobotany, use and possible future of yams in West Africa. Economic Botany 26 (4): 301-318.

BACO M. N., 2000. La domestication des ignames sauvages dans la sous-préfecture de Sinende: savoirs locaux et pratiques endogènes d'amélioration génétique des *Dioscorea abyssinica* Hochst. PhD thesis, Université nationale du Bénin, Faculté des sciences agronomiques, Cotonou, Benin, 170 p.

BAHUCHET S., 1982. Une société de chasseurs-cueilleurs et son milieu de vie: les Pygmées Aka de la forêt centrafricaine. PhD thesis, Institut des hautes études en sciences sociales, Paris, France, 616 p.

BAHUCHET S., MCKEY D., GARINE I., 1991. Wild yams revisited. Is independence from agriculture possible for rain forest hunter-gatherers? Human Ecology 19 (1): 213-243.

BAKER J. G., 1898. *Dioscoreaceae*. Flora of tropical Africa. W. T. Thiselton-Dyer, Lovell Reeve and Co, London, UK, 7 (3): 414-421.

BAQUAR S. R., 1980. Chromosome behaviour in Nigerian yams *(Dioscorea)*. Genetica 54:1-9.

BARALE G., LEMOIGNE Y., 2000. L'évolution de la flore. Pour la science (dossier La valse des espèces) 28: 36-41.

BINDER E., 1972. La génétique des populations. Collection Que sais-je? Presses universitaires de France, Paris, France, n° 1 283, 126 p.

BIO BIGOU L. B., 1994. Les origines du peuple Baatonou « Bariba », la société politique Wassangari et la question du trône de Nikki des origines à nos jours. Département de géographie, Université nationale du Bénin, Cotonou, Benin, 36 p.

BISSON P., 1989. Etude comparative des filières coton, riz, arachide et igname en savane ivoirienne. *In* M. GRIFFON Economie des filières en régions chaudes, formation des prix et échanges agricoles. Séminaire d'économie et de sociologie rurale des régions chaudes 10, 11-15 September 1989, Montpellier, France, 137 p. CIRAD, Montpellier, France. ISBN 2-87614-041-1.

BRICAS N., VERNIER P., ATEGBO E., HOUNHOUIGAN J., N'KPENU K. E., ORKWOR G. C., 1997. Le développement de la filière cossettes d'ignames en Afrique de l'Ouest. Cahiers de la Recherche Développement 44: 100-114.

BRICAS N., HOUNHOUIGAN J., 2000. Deux projets de recherche-développement. Bulletin du réseau TPA (Technologie et partenariat en agroalimentaire), dossier La transformation de l'igname, GRET, Paris, France, 18: 14-15. http://www.gret.org/tpa/bulletins/listebulletins.htm

BRICAS N., EL MOUSSAHOUI N., KAYODÉ P., NINDJIN C., ORKWOR G., HOUNHOUIGAN J., 2003. La consommation et les critères de qualité des ignames dans les villes du Bénin, de la Côte d'Ivoire et du Nigeria. Paper presented at the INCO international seminar Post-récolte et consommation des ignames, Cotonou, Benin, 17-19 June 2003.

BRUNET M., PICQ P., 2001. La grande expansion des Australopithèques. Des hominidés aux marges des forêts africaines. *In* Y. COPPENS and P. PICQ Aux origines de l'humanité, volume 1 De l'apparition de la vie à l'homme moderne, 652 p. Editions Fayard, Paris, France.

BURKILL I. H., 1918. Some cultivated yams from Africa and elsewhere. Gdns' Bull. Straits Settl. 2: 86-93.

BURKILL I. H., 1939. Notes on the genus *Dioscorea* in the Belgian Congo. Bull. Jard. Bot. Etat (Brussels) 15 (4): 345-392.

BURKILL I. H., PERRIER DE LA BATHIE H., 1950. *Dioscoreaceae* dans la flore de Madagascar et des Comores. Muséum national d'histoire naturelle, Paris, France, 78 p.

BURKILL I. H., 1960. The organography and the evolution of the *Dioscoreaceae,* the family of the yams. Botanical Journal of the Linnean Society (London) 56 (367): 319-420.

BURKILL H. M., 1985. The useful plants of west tropical Africa. Vol. 1 Families A-D, xvi + 960 p. Royal Botanic Gardens, Kew, UK. (Section *Dioscoreacea.* p. 654-670.)

CAILLIÉ R., 1932. Voyage de René Caillié à Tombouctou et à travers l'Afrique (1824-1928). Jacques Boulenger ed., Plon, Paris, France, in-12 paperback, XXIII-239 p.

CAMARA F., 2001. Structuration de la diversité génétique des cultivars d'ignames *Dioscorea cayenensis - rotundata* de la Haute Guinée. Analyse de la variabilité intravariétale par cytométrie en flux et Aflp. Agro.M, Université Montpellier II, Montpellier, France, 14 p.

CHAÏR H., AGBANGLA C., MARCHAND J. L., DAÏNOU O., NOYER J. L., 2005. Use of cpSSRs for the characterisation of yam phylogeny in Benin. Genome 48 (4). 674-684.

CHALINE J., 1999. Les horloges du vivant. Un nouveau stade de la théorie de l'évolution. Collection Sciences, Editions Hachette Littératures, Paris, France, 236 p.

CHEVALIER A., 1920. Exploration botanique de l'Afrique occidentale française. Tome 1. Editions Lechevallier, Paris, France, p. 639-643.

CHEVALIER A., 1936. Contribution à l'étude de quelques espèces africaines du genre *Dioscorea.* Bulletin du Muséum d'histoire naturelle, Paris, France, 2nd series 8 (6): 520-551.

CHEVALIER A., 1949. Sur une igname sauvage de l'Ouest africain à tubercules comestibles. Revue de Botanique Appliquée, p. 26-31.

CHIKWENDU V. E., OKEZIE C. E. A., 1989. Factors responsible for the ennoblement of African yams: inferences from experiments in yam domestication. Foraging and farming: the evolution of plant exploitation. *In* D. R. HARRIS and G. C. HILLMAN World Archeological Congress, Southampton, England, September 1986. Academic division of Unwin Hyman Ltd., London, UK, p. 344-357.

CHILAKA F. C., EZE S., ANYADIEGWU C., UVERE O., 2002. Browning in processed yams: Peroxidase or polyphenol oxidase? Department of Biochemistry, University of Nsukka, Nigeria, Journal of Food and Agriculture 82 (8): 899-903.

Coppens Y., Picq P., 2001. Aux origines de l'humanité. Volume 1: De l'apparition de la vie à l'homme moderne, 652 p. Volume 2: Le propre de l'homme, 576 p. Editions Fayard, Paris, France.

Conlan R. S., Griffiths L. A., Napier J. A., Shewry P. R., Mantell S., Ainsworth C., 1995. Isolation and characterisation of cDNA clones representing the genes encoding the major tuber storage protein (dioscorin) of yam (*Dioscorea cayenensis* Lam.). Plant Molecular Biology 28 (3): 369-380.

Coursey D. G., 1967. Yams. An account of the nature, origins, cultivation and utilisation of the useful members of the *Dioscoreaceae*. Tropical Agricultural Series, Longmans, Green and Co. Ltd, London, UK, 230 p.

Coursey D. G., 1976. The origins and domestication of yams in Africa. *In* J. R. Harlan, J. M. J. De Wet and A. B. L. Stemler The origins of African plant domestication, 498 p. Mouton Press, The Hague, Netherlands, p. 383-408.

Daïnou O., Agbangla C., Berthaud J., Tostain S., 2002. Le nombre chromosomique de base des espèces de *Dioscorea* constituant la section Enantiophyllum pourrait être égal à X = 20. Quelques preuves. Annales des sciences agronomiques du Bénin 3 (2), special edition.

Dansi A., 1995. Caractérisation des principaux cultivars du complexe *Dioscorea cayenensis - Dioscorea rotundata* du nord-est du Bénin. DEA graduate-level thesis, Université nationale de Côte d'Ivoire, Abidjan, 53 p.

Dansi A., Mignouna H. D., Zoundjihèkpon J., Sangare A., Asiedu R., Quin F. M., 1999. Morphological diversity, cultivar groups and possible descent in the cultivated yams *Dioscorea cayenensis-rotundata* complex in Benin Republic. Genetic Resources and Crop Evolution 46: 371-388.

Dansi A., Mignouna H. D., Zoundjihèkpon J., Sangare A., Ahoussou N., Asiedu R., 2000a. Identification of some Benin Republic's Guinea yam *(Dioscorea cayenensis-D. rotundata)* cultivars using Randomly Amplified Polymorphic DNA. Genetic Resources and Crop Evolution 47: 619-625.

Dansi A., Mignouna H. D., Zoundjihèkpon J., Sangare A, Asiedu R., Ahoussou N., 2000b. Using isozyme polymorphism for identifying and assessing genetic variation within cultivated yams (*Dioscorea cayenensis-Dioscorea rotundata* complex) of the Benin Republic. Genetic resources and Crop Evolution 47: 371-383.

Dansi A., Pillay M., Mignouna H. D., Daïnou O., Mondeil F., Moutaïrou K., 2000c. Ploidy level of the cultivated yams (*Dioscorea cayenensis-D. rotundata* complex) from Benin Republic as determinated by chromosome counting and flow cytometry. African Crop Science Journal (Uganda) 8 (4): 335-364.

Dansi A., Mignouna H. D., Pillay M., Zok S., 2001. Ploidy variation in the cultivated yams (*Dioscorea cayenensis-Dioscorea rotundata* complex) from Cameroon as determinated by flow cytometry. Euphytica 119: 301-307.

Degras L., Arnolin R., Poitout A., Suard C., 1977. Quelques aspects de la biologie des ignames (*Dioscorea* spp.). Les ignames et leur culture. Annales de l'amélioration des plantes 27 (1): 1-23.

DEGRAS L., 1986. L'igname: plante à tubercule tropicale. Collection Techniques agricoles et productions tropicales, Maisonneuve et Larose et ACCT, Paris, France, 408 p.

DOUMBIA S., 1998. Les déterminants agro-écologiques et socio-économiques de la production et de l'offre en igname en Côte-d'Ivoire. *In* Proceedings of the 12th Symposium of the International Society for Tropical Root Crops-Africa Branch (ISTRC-AB), Lilongwe, Malawi, 22-28 October 1995, 596 p. IITA, ISTRC-AB, Ibadan, Nigeria, p. 430-434.

DOUMBIA S., TSHIUNZA M., TOLLENS E., STESSENS J., 2004. Rapid spread of the Florido yam variety *(Dioscorea alata)*. Introduced for the wrong reasons and still a success. Outlook on Agriculture 33 (1): 49-54.

DOUNIAS E., 1996. Sauvage ou cultivé? La paraculture des ignames sauvages par les pygmées baka du Cameroun. *In* C. M. HLADIK, A. HLADIK, H. PAGEZY, O. F. LINARES, G. J. A. KOPPERT and A. FROMENT L'alimentation en forêt tropicale: interactions bioculturelles et perspectives de développement, 1 406 p. UNESCO Publishing, Paris, France, p. 939-960. ISBN 92-3-203381-X.

DOUNIAS E., 2001. The management of wild yam tubers by the Baka pygmees in Southern Cameroon. African Study Monographs 26: 135-156.

DUMONT R., 1971. L'igname au Nord Dahomey. 59 p. dactylographiées, IRAT; CIRAD, Montpellier, France.

DUMONT R., 1977. Etude morpho-botanique des ignames *Dioscorea rotundata* et *Dioscorea cayenensis* cultivées au Nord-Bénin. L'Agronomie Tropicale 32 (3): 225-241.

DUMONT R., 1980. Synthèse des études sur les tubercules de Haute-Volta. INERA, Burkina Faso; CIRAD, Montpellier, France, 87 pages.

DUMONT R., 1982. Ignames spontanées et cultivées au Bénin et en Haute-Volta. *In* J. MIÈGE and S. N. LYONGA Yams / Ignames Compte rendu de la Conférence internationale sur l'igname, International Foundation for Science (IFS), Buéa, Cameroon, 2-6 October 1978, 441 p. Clarendon Press, Oxford, UK, p. 31-36.

DUMONT R., 1988. Les stratégies de multiplication chez les ignames sauvages ouest-africaines. *In* L. DEGRAS Proceedings 7th Symposium ISTRC, Gosier, Guadeloupe, France, 1-6 July 1985. INRA, Versailles, France, p. 225-236.

DUMONT R., HAMON P., 1985. Une forme originale parmi les dioscoréacées cultivées en Afrique de l'Ouest. L'igname de Pilimpikou. Document CIRAD-IRAT; CIRAD, Montpellier, France, 6 p.

DUMONT R., 1993. Les ignames de Guinée. Aspects botaniques et techniques. Rapport de mission. Document CIRAD, IISDA-IITA; CIRAD, Montpellier, France, 23 p.

DUMONT R., HAMON P., SEIGNOBOS C., 1994. Les ignames du Cameroun. Collection Repères, CIRAD, Montpellier, France, 80 p.

DUMONT R., 1994. Conservation et évaluation des ressources génétiques à la station Idessa de Bouaké (Côte d'Ivoire). CIRAD, Montpellier, France, 4 p. + appendices

DUMONT R., 1995. Résultats d'une enquète conduite sur l'utilisation des cossettes d'ignames dans l'alimentation des populations de Cotonou (Bénin). Document CIRAD IITA, CIRAD, Montpellier, France, 8 p.

DUMONT R., 1997. La production d'ignames dans un village bariba du Bénin septentrional. Cahiers de la Recherche Développement 43: 35-51.

DUMONT R., VERNIER P., 1997a. La domestication des ignames *(D. cayenensis-rotundata)* chez l'ethnie bariba au Bénin. *In* Rencontre internationale, gestion des ressources génétiques des plantes en Afrique des savanes, 24-28 February 1997, Bamako, Mali, IER, BRG, Solagral, 352 p. IER, Sikasso, Mali; BRG, Paris, France; Solagral, Nogent-sur-Marne, France, p. 47-54.

DUMONT R., VERNIER P., 1997b. L'igname en Afrique: des solutions transférables pour le développement. Cahiers de la Recherche Développement 44: 115-120.

DUMONT R., VERNIER P., 2000. Domestication of yams *(Dioscorea cayenensis - rotundata)* within the Bariba ethnic group in Benin. Outlook in Agriculture 29 (2): 137-142.

DUMONT R., KOUAKOU A.M., *(submitted)*. Influence de la densité de plantation et de la fertilisation chimique sur la production commerciale et semencière des ignames *Dioscorea rotundata* Poir. à deux récoltes cultivées dans la zone de savane de Côte d'Ivoire.

DURAND T., DURAND H., 1909. Syll. Fl. Congol., p. 568-561.

EDEOGA H. O., OKOLI B. E., 2001. Mid-rib anatomy and systematics in *Dioscorea* L. *(Dioscoreaceae)*. Journal of Economic and Taxonomy Botany, Additional Series 19: 191-195.

ESSAD S., 1984. Variation géographique des nombres chromosomiques de base et polyploïdie dans le genre *Dioscorea,* à propos du dénombrement des espèces *transversa* Brown, *pilosiuscula* Bert et *trifida* L. Agronomie 4: 611-617.

FAOSTAT, 2003. FAO online statistic database: http://apps.fao.org. Queried 4 April 2005.

FAUTRET A., 1985. Callogénèse et néoformation chez deux espèces d'ignames comestibles: *Dioscorea alata* et *D. trifida.* PhD thesis, Université de Montpellier II, Montpellier, France, 163 p.

GAMIETTE F., BAKRY F., ANO G., 1999. Ploidy determination of some yam species *(Dioscorea* spp.) by flow cytometry and conventional chromosomes counting. Genetic Resources and Crop Evolution 46: 19-27.

GEBRE MARIAM T., SCHMIDT P. C., 1996. The use of a starch obtained from *Dioscorea abyssinica* in tablet formulations. I. The native starch as a binder and disintegrant. Pharmazeutische Industrie 58: 167-172.

GENETIC RESOURCE ACTION INTERNATIONAL (GRAIN), 2002. Les brevets et la diversité biologique africaine. Grain, Barcelona, Spain, Semences de la biodiversité n° 10.

GHARTEY A.B., 1994. Export of yams from Ghana. Report to the Post-Harvest Project. MOFA, Ministry of Food and Agriculture, GTZ Technical Cooperation, Accra, Ghana, 55 p.

GOUDOU-URBINO C., KONATE G., QUIOT J. B., DUBERN J., 1996. Ethiology and ecology of a yam mosaic disease in Burkina Faso. Tropical Science 36 (1): 34-40.

GOULD S. J., HALL B. K., 1999. *In* J. CHALINE Les horloges du vivant. Un nouveau stade de la théorie de l'évolution, 236 p. Editions Hachette Littératures, Paris, France.

GOUYON P. H., HENRY J. P., ARNOULD J., 1997. Les avatars du gène. La théorie néodarwinienne de l'évolution. Collection Regards sur la science, Editions Belin, Paris, France, 336 p.

HALLÉ F., 1999. Eloge de la plante. Pour une nouvelle biologie. Editions du Seuil, Paris, France, 329 p.

HAMON P., 1987. Structure, origine génétique des ignames cultivées du complexe *Dioscorea cayenensis - rotundata* et domestication des ignames en Afrique de l'Ouest. PhD thesis, Université Paris XI, Orsay, France, 247 p.

HAMON P., BRIZARD J. P., ZOUNDJIHÈKPON J., DUPERRAY C., BORGEL A., 1992. Etude des index d'ADN de huit espèces d'ignames (*Dioscorea* sp.) par cytométrie en flux. Canadian Journal of Botany 70 (5): 996-1 000.

HAMON P., DUMONT R., ZOUNDJIHÈKPON J., TIO-TOURÉ B., HAMON S., 1995. Les ignames sauvages d'Afrique de l'Ouest. Collection Didactiques, Editions de l'Orstom, IRD, Bondy, France, 84 p.

HAMON P., TIO-TOURÉ B., 1982. Etude du polymorphisme enzymatique par électrophorèse sur gel d'amidon de quelques populations d'ignames spontanées et cultivées de Côte d'Ivoire. Annales de l'Université d'Abidjan, Côte d'Ivoire, Série C (Sciences) 18: 99-112.

HARTMANN C., JOSEPH C. MILLET B., 1998. Age chronologique, âge physiologique et activités rythmiques. Biologie et physiologie de la plante. Collection Fac/Sciences, Editions Nathan, Paris, France, 224 p. ISBN 2-02-190939-4.

HILDEBRAND E. A., 2003. Motives and opportunities for domestication and ethnoarchaeological study in southwest Ethiopa. Journal of Anthropological Archaeology 22: 358-375.

HLADIK A., BAHUCHET S., DUCATILLON C., HLADIK C. M., 1984. Les plantes à tubercules de la forêt dense d'Afrique centrale. Revue d'Ecologie la Terre et la Vie (Paris, France) 39: 249-290.

HLADIK A., DOUNIAS E., 1996. Les ignames spontanées des forêts denses africaines, plantes à tubercules comestibles. L'alimentation en forêt tropicale: interactions bioculturelles et perspectives de développement. *In* C. M. HLADIK, A. HLADIK, H. PAGEZY, O. F. LINARES, G. J. A. KOPPERT and A. FROMENT L'alimentation en forêt tropicale: interactions bioculturelles et perspectives de développement, 1 406 p. UNESCO Publishing, Paris, France, p. 275-294. ISBN 92-3-203381-X.

HLADIK C. M., 1985. *In* Chapter 21, discussion. *In* COPPENS Y. *et al.* L'environnement des Hominidés au plio-pleistocène, 468 p. Fondation Singer-Polignac, Editions Masson, Paris, France, p. 447-452.

HUTCHINSON J. DALZIEL J. M., 1931. Flora of tropical West Africa. Vol. II: p. 373-382. Crown Agents, London.

IDESSA, 1986. Institut des savanes. Centre de recherche sur les cultures vivrières. Rapport d'activité du programme ignames, campagne 1986. Bouaké, Côte d'Ivoire.

INRAB, 1996. Synthèse des rendements enregistrés sur la collection d'ignames de 1982 à 1995. Institut de la recherche agronomique du Bénin, Station d'INA (unpublished document).

IGUÉ J., 1985. Rente pétrolière et commerce des produits agricoles à la périphérie du Nigeria: les cas du Bénin et du Niger. INRA / ESR-LEI, Montpellier, France, 103 p.

IWU M. M., OKUNJI C. O., AKAH P., TEMPESTA M. S., CORLEY D., 1990. Dioscoretane: the hypoglycemic principle of *Dioscorea dumetorum*. Planta Medica 56: 119-121.

JACQUES-FÉLIX H., 1947. Ignames sauvages et cultivées du Cameroun. Revue de Botanique Appliquée 27 (293-294): 119-133.

JAEGER J. J., 2001. La terre avant les hommes. *In* Y. COPPENS and P. PICQ Aux origines de l'humanité, volume 1 De l'apparition de la vie à l'homme moderne, 652 p. Editions Fayard, Paris, France, p. 25-71.

KASSAMADA K., 1992. Contribution à la caractérisation morphologique des cultivars du complexe *Dioscorea cayenensis-rotundata*. PhD thesis, Université du Bénin; Ecole supérieure d'agronomie, Lomé, Togo, 111 p.

KAYODÉ A., HOUNHOUIGAN J., MESTRE C., AKISSOE N., MEOT J., 2005. Influence de différentes conditions de pré-cuisson sur la qualité des farines et pâtes de cossettes d'igname *(Dioscorea cayenensis - rotundata)*. Annales des sciences agronomiques du Bénin.

ILU O. Y. T., 1993. Induced dwarf-type mutant of yam *Dioscorea rotundata* Poir. Tropical Agriculture 70 (3): 289-290.

KOUROUMA S., 2001. Yam cultivation in Guinea. *In* West Africa seed and Planting Material, Newsletter of the West Africa Seed Network (WASNET) n° 8, 31 p., October 2001, p. 24-26. West Africa Seed Development Unit (WASDU), PO Box 9698, K.I.A., Accra, Ghana. [www.wasnet.org/newsletter/english/WASNET8%20English.pdf].

KUPZOW A. J., 1980. Theoretical basis of the plant domestication. Theoretical and Applied Genetics 57 (2): 65-74.

LAWTON J. R. S., LAWTON J. R., 1967. The morphology of the dormant embryo and young seedling of five species *Dioscorea* from Nigeria. Proc. Linn. Soc. 178 (2): 153-159.

LÉZINE A. M., 2000. L'épisode froid du Dryas. Pour la science, French edition of *Scientific American,* special edition La valse des espèces, p. 128-131.

LI P., JOHNSTON M. O., 2000. Heterochrony in plant evolution studies though the twentieth century. Botanical Review 66 (1): 57-88.

LOMBARD J., 1965. Structures de type féodal en Afrique Noire, études des dynamismes internes et des relations sociales chez les Baribas du Dahomey. Editions Mouton, Paris, France.

LYONGA S. N., 1976. Production investigations on edible yams (*Discorea* spp.) in the western and north western highland savannah zone of the United Republic of Cameroon. PhD thesis, Université d'Ibadan, Nigeria, 298 p.

MANGENOT G., 1955. Ecologie et représentation cartographique des forêts équatoriales et tropicales humides. Annales biologiques (Paris) 31: 149-156.

MARTIN F. W., ORTIZ S., 1963. Chromosome numbers and behaviour in some species of *Dioscorea*. Cytologia 28: 96-101.

MARTIN F. W., 1969. The species of *Dioscorea* containing sapogenin. Economic Botany 23 (4) 373-379.

MARTIN F.W. SADIK S., 1977. Tropical yams and their potential. Part. 4. *Dioscorea rotundata* and *Dioscorea cayenensis*. US Department of Agriculture, Agriculture Handbook n° 502.

MANYONG V. M., SMITH J., WEBER G., JAGTAP S. S., OYEWOLE B., 1996. Macro-characterization of agricultural systems in West Africa. An overview. Resource and Crop Management Division, IITA, Ibadan, Nigeria, Research Monograph n° 21, 66 p.

MBAILO KEMDINGAO L., 1998. Recherche sur la culture de l'igname au Tchad. *In* J. BERTHAUD, N. BRICAS and J. L. MARCHAND L'igname, plante séculaire et culture d'avenir. International seminar proceedings CIRAD, INRA, ORSTOM, CORAF, 3-6 June 1997, Montpellier, France, 453 p. CIRAD, Montpellier, France, p. 397-398. ISBN 2-87614-313-5.

MIÈGE J., 1952. Contribution à l'étude systématique des *Dioscorea* d'Afrique occidentale. Thèse, faculté des sciences, Paris, France, 266 p.

MIÈGE J., 1954. Nombres chromosomiques et répartition géographique de quelques plantes tropicales et équatoriales. Revue de Cytologie et de Biologie Végétale, volume XV, part 4.

MIÈGE J., 1958. Deux ignames nouvelles d'Afrique occidentale à tubercules vivaces. Bulletin de l'IFAN, Institut fondamental d'Afrique Noire Cheikh Anta Diop, Senegal, 20 (1).

MIÈGE J., 1968. Dioscoreaceae. *In* J. HUTCHINSON, J. M. DALZIEL and F. N. HEPPER Flora of tropical West Africa, 2. London, Crown Agents, 3: 144-154.

MIÈGE J., 1979. Sur quelques problèmes taxonomiques posés par *Dioscorea cayenensis* Lam. et *D. rotundata* Poir. Taxonomic aspects of African Economic Botany. *In* Proceedings of the 9th Plenary Meeting of AETFAT, Las Pamas de Gran Canaria. Las Palmas, Gran Canaria, Kunkel, p. 109-113.

MIÈGE J., LYONGA S. N., 1982. Yams/Ignames. Compte rendu de la Conférence internationale sur l'igname, International Foundation for Science (IFS), Buéa, Cameroon, 2-6 October 1978. Clarendon Press, Oxford, UK, 441 p.

MIGNOUNA H. D., ASIEDU R., THOTTAPPILLY G., 1998. Molecular taxonomy of cultivated and wild yams in West Africa. *In* M. O. AKORODA and I. J. EKANAYAKE Proceedings 6th Symposium of the International Society for Tropical Root Crops-Africa Branch (ISTRC-AB), Lilongwe, Malawi, 22-28 October 1995. ISTRC-AB, IITA, Ibadan, Nigeria, p. 429.

MIGNOUNA H. D., DANSI A., 2002. Yam (*Dioscorea* sp.) domestication with the Nago and Fon ethnic group of Benin Republic. Genetic Resources and Crop Evolution 50: 519-528.

MIGNOUNA H. D., DANSI A., 2003. 6-Phosphogluconate dehydrogenase (6-PGD) in yam (*Dioscorea* spp.): variation and classification. Plant Genetic Resources Newsletter 133: 27-30.

MIGNOUNA H. D., DANSI A., ZOK S., 2002a. Morphological and isozymic diversity of the cultivated yams (*Dioscorea cayenensis-Dioscorea rotundata* complex) of Cameroon. Genetic Resources and Crop Evolution 49: 21-29.

MIGNOUNA H. D., MANK R. A., ELLIS T. H. N., VAN DEN BOSCH N., ASIEDU R., NG S. Y.C., PELEMAN J., 2002b. A genetic linkage map of Guinea yam (*Dioscorea rotundata* Poir.) based on AFLP markers. Theoretical and Applied Genetics 105 (5): 716-725.

NAIR S. G., ABRAHAM K. A., 1992. Dwarfing and its potential in the improvment of white yam. Journal of Root Crops 18 (2): 136-138.

N'KOUNKOU J. S., 1993. La section Enantiophyllum Uline du genre *Dioscorea* L. en Afrique centrale. Belgian Journal of Botany 126 (1): 435-470.

N'KOUNKOU J. S., 1994. Analyses phylogénétiques des *Dioscorea (Dioscoreaceae)* d'Afrique centrale (Congo, Zaïre, Rwanda, Burundi). Fragmenta Floristica Geobotanica 39 (2): 401-416.

OKEZIE C. E. A., OKONKWO S. N., NWOKE F. I., 1993. The effect of daylength on growth and tuberisation of *Dioscorea rotundata* propagated by seed. Nigerian Journal of Botany 6: 53-59.

OKRY K. F., 2000. L'igname dans le système de production agricole de Banté et la domestication de quelques-unes de ses formes sauvages: savoirs locaux et pratiques endogènes de culture et d'amélioration génétique. PhD thesis, Université nationale du Bénin, Faculté des sciences agronomiques, Cotonou, Benin, 87 p.

ONWUEME I. C., CHARLES W. B., 1994. Tropical root and tuber crops. Production, perspective and future prospects. FAO, Rome, Italy, FAO Plant Production and Protection Paper n° 126, 112 p.

ONYILAGHA J. C., LOWE J., 1985. Studies on the relationships of *Dioscorea cayenensis* and *Dioscorea rotundata* cultivars. Euphytica 35: 733-739.

OSAGIE A. U., 1992. The yam tuber in storage. An up-to-date review of the biochemical composition and storage of the yam tuber. Published by Postharvest Research Unit, Department of Biochemistry, University of Benin, Benin City, Edo State, Nigeria, 247 p.

PÉRON Y., ZALACAIN V., 1975. Atlas de Haute-Volta. Editions Jeune Afrique, Paris, France, 47 p.

PFEIFFER H. I., 1987. From small scale farmers contraints to farmers adapted improvements. CNRCIP Adamaoua, Cameroon, 35 p.

PIPERNO D. R., HOLST I., 1998. The presence of starch grains on prehistoric stone tools from the humid neotropics. Indications of early tuber use and agriculture in Panama. Journal of Archaelogical Science 25 (8): 765-776.

PITKIN B. R., 1973. Larothrips *(Thysanoptera, thripidae)* a new genus and species of thrips from yam flowers in Nigeria. Bulletin of Entomological Research 62: 415-418.

POETHIG X., SCOTT R., 1998. Timing is everything: genetic and temporal regulation of leaf identity. *In* Session 49 Symposium Hetechrony in plants, Annual Meeting of the Botanical Society of America, 2-6 August, 1998, Baltimore, Maryland, USA.

RAADTS E., 1984. Dioscoreaceae. *In* J.-F. BRUNEL, P. HIEPKO and H. SCHOLZ Flore ana-lytique du Togo. Phanérogames. Deutsche Gesellschaft für Technische Zusammenarbeit (Gtz) gmbH, Eschborn, Germany, p. 567-570.

RAMSER J., WEISING K., LOPEZ-PERALTA C., TERHALLE W., TERAUCHI R., KAHL G., 1997. Molecular marker based taxonomy and phylogeny of guinea yam *(D. rotundata - cayenensis)*. Genome 40 (6): 903-915.

ROUGERIE G., 1982. La Côte d'Ivoire. Collection Que sais-je? 5th edition. Presses uni-versitaires de France, Paris, France, 128 p.

SADIK S., OKEREKE O. U., 1975. Flowering, pollen grain germination, fruiting, seed germination and seedling development of white yam, *Dioscorea rotundata*. Annals of Botany 39: 597-604.

SCARCELLI N., 2002. Etude des relations entre les ignames cultivées et les formes sauvages apparentées (*Dioscorea* sp.) au Bénin. Impact de la domestication sur la diver-sité des plantes cultivées. DEA graduate-level thesis, Agro.M, Université Montpellier II, Montpellier, France, 46 p.

SCARCELLI N., TOSTAIN S., MARIAC C., AGBANGLA C., DA O., BERTHAUD J., PHAM J. L., 2006. Genetic nature of yams (*Dioscorea* sp.) domesticated by farmers in Benin (West Africa). Genetic Resources and Crop Evolution 53 (1): 121-130.

SCHNELL R., 1971. Introduction à la phytogéographie des pays tropicaux. Volume 2. Gauthier-Villars Editeurs, Paris, France, 447 p.

SCHOLS P., FURNESS C. A., WILKIN P., HUYSMANS S., SMETS E., 2001. Morphology of pollen and orbicules in some *Dioscorea* species and its systematic implications. Botanical Journal of the Linnean Society 136 (3): 295-311.

SCOTT G., ROSEGRANT M., RINGLER C., 2000. Roots and tubers for the 21th century. DP 31, IFPRI-CIP, 71 p.

SEIGNOBOS C., 1998. Evolution d'un agrosystème à ignames: l'exemple des Dorou du Nord-Cameroun. *In* J. BERTHAUD, N. BRICAS and J.-L. MARCHAND L'igname, plante séculaire et culture d'avenir. International seminar proceedings CIRAD, INRA, ORSTOM, CORAF, 3-6 June 1997, Montpellier, France, 453 p. CIRAD, Montpellier, France, p. 51-57. ISBN 2-87614-313-5.

SEIGNOBOS C., 1992. L'igname dans les monts Mandara (Nord-Cameroun). Genève Afrique 30 (1): 78-96.

SENIOU D., 1993. Contribution à la caractérisation par électrophorèse enzymatique des cultivars du complexe *Dioscorea cayenensis - rotundata*. PhD thesis, Ecole supérieure d'agronomie du Togo, Lomé, Togo, 103 p.

SHIWACHI H., AYANKANMI T., ASIEDU R., 2003. A technique for grafting of water yam *(Dioscorea alata)*. *In* M. AKORODA Proceedings of 8th Symposium of the International Society for Tropical Root Crops-Africa Branch (ISTRC-AB), Ibadan, Nigeria, 12-16 November 2001. IITA, ISTRC-AB, Ibadan, Nigeria, p. 265-267.

SOURNIA J. C., 1991. Dictionnaire de génétique. Edition OLF Sa, diffusion CILF, Paris, France, 351 p. ISBN 2-85319-231-8.

TERAUCHI R., CHIKALEKE V. A., THOTTAPPILLY G., HAHN S. K., 1992. Origin and phylogeny of Guinea yams as revealed by RFLP analysis of chloroplast DNA and nuclear ribosomal DNA. Theoretical Applied Genetics 83 (6/7): 743-751.

TRÈCHE S., 1989. Potentialités nutritionnelles des ignames *(Dioscorea* spp.) cultivées au Cameroun. Vol. I: text. Vol. II: appendices. Collection Etudes et Thèses, IRD, Bondy, France, 595 p.

TOKPA G., DUMONT R., 1995. Les cultivars d'ignames cultivées dans la région de Dikodougou. *In* Document de travail n° 3. Renforcement des études agro-économiques à l'Idessa. Institut des Savanes, Bouaké, Côte d'Ivoire; Université catholique de Louvain, Belgium, 26 p.

TOSTAIN S., 1998. A bibliography on yams and their utilization, 1984-1996: *In* J. BERTHAUD, N. BRICAS and J.-L. MARCHAND L'igname, plante séculaire et culture d'avenir. International seminar proceedings CIRAD, INRA, ORSTOM, CORAF, 3-6 June 1997, Montpellier, France, 453 p. CIRAD, Montpellier, France. ISBN 2-87614-313-5.

TOSTAIN S., DAÏNOU O., 1998. Bilan des collectes de *Dioscorea abyssinica* et de *D. praehensilis* en 1996 et 1997; première étape d'un programme d'étude des relations entre ignames sauvages et ignames cultivées au Bénin. *In* J. BERTHAUD, N. BRICAS and J.-L. MARCHAND L'igname, plante séculaire et culture d'avenir. International seminar proceedings CIRAD, INRA, ORSTOM, CORAF, 3-6 June 1997, Montpellier, France, 453 p. CIRAD, Montpellier, France, p. 257-258. ISBN 2-87614-313-5.

TOSTAIN S., AGBANGLA C., DAÏNOU O., 2002. Les ignames *Dioscorea abyssinica* et *D. praehensilis* en Afrique de l'Ouest. Diversité génétique estimée par des marqueurs Aflp. Annales des Sciences Agronomiques du Bénin 3: 1-20.

TOURÉ B., AHOUSSOU N., 1982. Etude du comportement en collection des ignames *(Dioscorea* spp) dans deux régions écologiques différentes de la Côte d'Ivoire. *In* J. MIÈGE and S.N. LYONGA Yams / Ignames, compte rendu de la Conférence internationale sur l'igname, International Foundation for Science (IFS), Buéa, Cameroon, 2-6 October 1978, 441 p. Clarendon Press, Oxford, UK, p. 23-30.

TOURÉ M., STESSENS J., MAYAO A. G., ZOHOURI G. P., 2003. Variation saisonnière de l'offre et des prix de l'igname en Côte d'Ivoire. Agronomie Africaine (special edition Atelier national sur l'igname d'octobre 2001) 4: 69-82.

TRÈCHE S., 1989. Potentialités nutritionnelles des ignames *(Dioscorea* spp.) cultivées au Cameroun. Vol I (text), vol II (appendices). Collection Etudes et Thèses, IRD, Bondy, France, 595 p.

VALDEYRON G., 1961. Génétique et amélioration des plantes. Editions Baillère et Fils, Paris, France, 374 p.

VANDEVENNE R., CASTANIÉ D., 1988. Croissance et développement de trois cultivars d'ignames appartenant aux espèces *Dioscorea cayenensis, D. rotundata* et *D. alata* en Côte d'Ivoire. *In* L. DEGRAS Proceedings 7th Symposium of the International Society for Tropical Root Crops (ISTRC), Gosier (Guadeloupe), 1-6 July 1985, 939 p. INRA, Versailles, France, p. 201-219. ISBN 2-7380-0029-0.

VERNIER P., N'KPENU K. E., ORKWOR G. C., 1999. La demande urbaine en cossettes d'igname. Conséquences sur la filière de production d'ignames. Agriculture et développement 23: 32-43.

VERNIER P., DANSI A., 2000. Participatory assessment and farmer knowledge on yam varieties *(D. rotundata)* in Benin. Communication *In* M. NAKATANI and K. KOMAKI Potential of root crops for food and industrial resources, Proceedings 12th Symposium of the International Society for Tropical Root Crops (ISTRC), 10-18 September 2000, Tsukuba, Japan, 611 p. ISTRC, p. 360-365.

VERNIER P., DOSSOU A. R., 2000. Adaptation of yam (*Dioscorea* spp.) cultivation to changing environment and economic constraints in Benin, West Africa. Communication *In* M. NAKATANI and K. KOMAKI Potential of root crops for food and industrial resources, Proceedings 12th Symposium of the International Society for Tropical Root Crops (ISTRC), 10-18 September 2000, Tsukuba, Japan, 611 p. ISTRC, p. 352-359.

VERNIER P., ORKWOR G. C., DOSSOU A. R., 2003. Studies on yam domestication and farmers' practices in Benin and Nigeria. Outlook on Agriculture 32 (1): 35-41.

WILKIN P., 2001. *Dioscoreaceae* of South-Central Africa. Kew Bulletin 56 (2): 361-404.

WAITT A. W., 1965. A key to some Nigerian varieties of yam (*Dioscorea* spp.). Fred. Dept. Agric. Res., Nigeria, Memorendum 60.

ZOUNDJIHÈKPON J., ESSAD S., TIO-TOURÉ B, 1990. Dénombrement chromosomique dans dix groupes variétaux du complexe *Dioscorea cayenensis - rotundata*. Cytologia 55 (1): 115-120.

ZOUNDJIHÈKPON J., 1993. Biologie de la reproduction et génétique des ignames cultivées de l'Afrique de l'Ouest, *Dioscorea cayenensis - rotundata*. PhD thesis n° 194, Université nationale de Côte d'Ivoire, Faculté des sciences et techniques, Abidjan, Côte d'Ivoire, 305 p.

ZOUNDJIHÈKPON J., HAMON P., TIO-TOURÉ B., HAMON S., 1994. First controlled progenies checked by isozymic markers in cultivated yams *Dioscorea cayenensis-rotundata*. Theoretical Applied Genetics 88: 1 011-1 016.

Layout: Isabelle Corniquet, Montpellier (France)
Fichier préparé par Nicolas Perrier, société 4P
Imprimé pour vous par Libri Plureos GmbH (Allemagne)